NOW 2 kNOW!

Geometry

by T. G. D'Alberto

Pithy Professor Publishing Company
Brighton, CO

Published by

Pithy Professor Publishing Company, LLC
PO Box 33824
Northglenn, CO 80233

ISBN: 978-0-9882054-5-1

Library of Congress Control Number: 2014913675

Printed in the United States of America

About the Author

Dr. Tiffanie G. D'Alberto has a Ph.D. in Electrical & Computer Engineering from Cornell University and a B.S. and M.S. in Electrical Engineering from Virginia Polytechnic Institute & State University.

She has worked for over a decade in the telecommunications and aerospace industries as a scientist, program manager, and supervisor. She has engaged in numerous opportunities for tutoring, teaching, and mentoring throughout her career and schooling.

In her spare time, Tiffanie enjoys oil painting, drawing, reading, sewing, and running. She's a huge fan of Star Trek, Renaissance Festivals, and animals.

Tiffanie lives in St. Croix with her fiancé, Colin, and their many wonderful pets.

Dedication

To my dearest Colin, who inspires me, encourages me, and supports me. I could never thank you enough.

To my high school Geometry teacher, Mr. Rihard, who made learning this subject both easy and fun.

Acknowledgements

I always thank my family first: My parents for the foundation, the push, and the belief in me all along; My fiancé for his inspiration, encouragement, and unending support.

A huge thanks goes to my many excellent math teachers from middle school to high school to college that not only taught the material but also taught the way of thinking necessary to excel in these subjects.

I'd also like to acknowledge Barron's Review Course Series **Let's Review: Geometry** by Lawrence S. Leff, M.S. for a thorough refresher of the material.

Finally, I'd like to thank Amazon.com for their excellent publish-on-demand service that enables books such as these, and you, the reader, for making this investment in your future.

Table of Contents

Introduction

Welcome!

Geometry is the study of objects & shapes. It is heavy on definitions and proofs of theorems. Because of this, the first two chapters of this book are <u>BORING</u>. I could waste a bunch of pages, and thus your time, by trying to make the basics interesting, but this book is about getting you information quickly. Never fear, we actually get to do Geometry in Chapter 3, and many students find it a nice break from the usual math course.

The process of learning Geometry and math is three-fold:

1. **To excel at math is to understand math.** For example, you know how to play Go Fish. You not only know the rules, you understand the object of the game and the techniques that are required to dominate against your 4-year old opponent. It doesn't matter that this time you have a different hand or that's it's been 10 years since you last played. You *understand* the game, so you can play it well. That's how you should learn math.

2. **To understand math, you need the story** . The story is the logic flow that allows you to keep building on your *understanding*. If someone tells you a story and skips a critical part of the plot, you would and should say, "Hey, back up!"

3. **To understand math, you also need the big picture.** The big picture is the outline of the logic, or story, placed in an area small enough for you to see it in its entirety. Like a file directory on a computer, it organizes the information. Once you see the flow of the big picture, it's easier for you to put the details of the story into their proper places.

The key to learning math is not memorization, it's understanding. Be open to changing the way you think. Once you get the flow, you'll get the A's. I wish you great success!

Layout:

The layout of this text is different from most academic books:

1. **The problem sets are saved to the end of the book.** In this book, you can read from beginning to end to understand the logical progression of the course, or stop to do problem sets as you desire.

2. **Solution sets give the critical steps to get the answers, not just the answers.** Because this is not a textbook for a classroom, there is no need to keep the "secret sauce" from you. Use the problems as drills or study the solutions as further examples.

3. **Appendix A is an overall summary of the entire book.** It helps you visualize the big picture and logic flow to give you a framework into which you can organize the details.

In addition, the following visual markers will help you navigate the material...

Key terms defined for the first time are **bolded** and also found in the index.

Important equations are shown as:

> *important information*

Finally, examples are given as supplements to the text as well as for illustration:

> *Example* — This is an example to illustrate a point or to give further definition. Skip it if you feel very comfortable with the material presented thus far.

Illustrative graphics and additional notes are shown on the side to accompany the text.

Chapter 1: The Basics

Definitions:

As mentioned in the introduction, there are a lot of definitions in Geometry. Let's get started:

Point: An infinitely small point in space shown with a dot. Named with a capital letter. Ex: point A

Coordinate: A pair (or triplet) of numbers giving the x, y, (and z) locations of a point. Ex: (1,2); (2,3,4).

Collinear points: Points on the same line are collinear.

Line: A straight connection between two points that extends from $-\infty$ to $+\infty$. It is named by the points it connects with a double arrow over the top. Ex: $\overleftrightarrow{AB}$

Line segment: A portion of a line. Named like a line but with a line segment over top. Ex: $\overline{AB}$

Ray: A line that extends infinitely in one direction. Named like a line but with a ray over top. Ex: $\overrightarrow{AB}$

Slope: The rise over the run of a line indicating where the line is pointing from left to right:

$$m = \frac{\Delta y}{\Delta x} = \frac{y_2 - y_1}{x_2 - x_1}$$

Parallel: Lines, segments, or rays in the same plane that never intersect; $m_{AB} = m_{CD}$. Notation: $\overleftrightarrow{AB} \parallel \overleftrightarrow{CD}$

Perpendicular: Intersecting lines, segments, or rays that are square to each other; $m_{CD} = \frac{-1}{m_{EC}}$. Notation: $\overleftrightarrow{CD} \perp \overline{EC}$

Skew: Two lines, segments, or rays are skew if they are non-parallel and also do not intersect. The two lines are in different planes.

Plane: A plane is an infinitesimally thin, flat surface that extends infinitely in all directions. Notation is the same as a point. Ex: plane P

Coplanar: Two lines are coplanar if they are in the same plane.

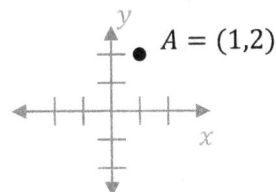

Point A.

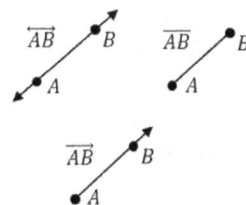

Line, segment, ray.

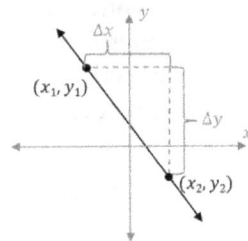

Slope.

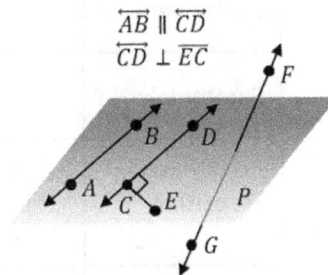

Plane P: $\overleftrightarrow{FG}$ is skew to all other lines and is noncoplanar.

Angle: An angle is formed by the intersection of two lines (or segments or rays). Named with an L-shaped symbol followed by either one or three point designations. Ex: $\angle B$; $\angle ABC$

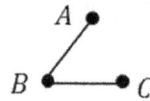

Angle with vertex at point B.

Vertex: The vertex of the angle is the point where the two lines intersect. It is used as the single or middle letter in the naming scheme.

Acute: An angle is acute if it is less then square.

Right: An angle is right if it is square. A square is usually placed at the vertex to show a right angle.

Obtuse: An angle is obtuse if it is greater than square.

Straight: An angle is straight if it forms a line.

Acute, Right, Obtuse, and Straight angles.

Adjacent: Two angles are adjacent if they share a side.

Complement: Two angles are complementary if they add to form a right angle.

Supplement: Two angles are supplementary if they add to form a straight line.

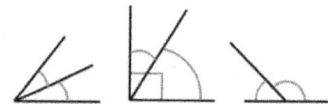

Adjacent, Complimentary, and Supplementary angles.

Congruent: Two line segments or two angles are congruent if their measures are the same. Matching tic marks on a figure show which quantities are equal. We also have two equivalent ways to write congruency :

$$AB = DE \quad \text{or} \quad \overline{AB} \cong \overline{DE}$$
$$m\angle B = m\angle E \quad \text{or} \quad \angle B \cong \angle E$$

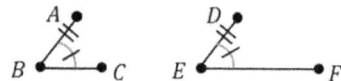

Congruent lines and angles.

Properties of Congruency: Congruency follows these rules:

Reflexive	Things are congruent to themselves	$\overline{AB} \cong \overline{AB}$
Symmetric	Order doesn't matter	If $\overline{AB} \cong \overline{CD}$ Then $\overline{CD} \cong \overline{AB}$
Transitive	If two things equal a third thing, then the two things are equal to each other	If $\overline{AB} \cong \overline{EF}$ And $\overline{CD} \cong \overline{EF}$ Then $\overline{AB} \cong \overline{CD}$
Substitution	A thing can be substituted with its equal	If $m\angle A + m\angle B = x$ And $\angle B \cong \angle C$ Then $m\angle A + m\angle C = x$

Tools of the Trade:

There are three main tools of the trade: a ruler, a protractor, and a compass.

Ruler: Rulers measure the **distance** between points (the length of line segments), and it can help locate the **midpoint** (center) between two points. Recall from Algebra 1 that if you have the coordinates of point 1 (x_1, y_1) and point 2 (x_2, y_2), the distance and midpoint are determined as follows:

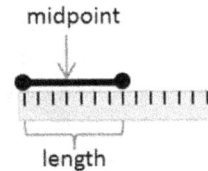

midpoint

length

A ruler.

$$distance = d = \sqrt{(x_2 - x_1)^2 + (y_2 - y_1)^2}$$

$$midpoint = \left(\frac{x_1+x_2}{2}, \frac{y_1+y_2}{2}\right)$$

Protractor: The protractor measures angles. You put the straight edge along one side of the angel with its vertex centered, then read the angle measurement on the arc where the other side crosses.

A protractor.

Angles are measured as portions of a whole circle in either degrees or radians.

- **Degrees:** There are 360 degrees (360°) in a full circle, so a straight angle that cuts the circle in half is 180°, and a right angle, or ¼ of a circle, is 90°.

A straight angle and right angle is ½ and ¼ of a circle, respectively.

- **Radians:** There are 2π radians in a circle, so a straight angle is π radians (or rad), and a right angle is $\frac{\pi}{2}$ rad. The value of π is ~3.14159 and represents the ratio of any circle's circumference to its diameter (see Chapter 10 for definitions).

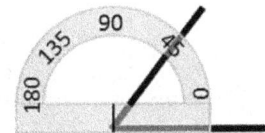

$$45° = 45° \times \frac{\pi}{180°} = \frac{\pi}{4}$$

- **Degree to Radian Conversion:** Since 180° $= \pi$, to convert between degrees and radians multiply or divide by 180°/π to get the correct units.

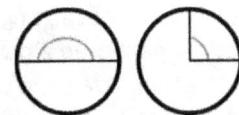

$$\frac{\pi}{3} = \frac{\pi}{3} \times \frac{180°}{\pi} = 60°$$

Degree-Radian Conversion.

Compass: A compass is ideal for drawing circles and arcs, but it also draws bisectors. A **bisector** is a line that splits a line segment or angle in half.

To draw a perpendicular bisector for a line:
1. Place the compass point on one endpoint and draw an arc that intersects the line
2. Do the same with the compass point on the other endpoint.
3. Place a point on each of the two intersections of the arcs.
4. Draw a line between the points.

A compass.

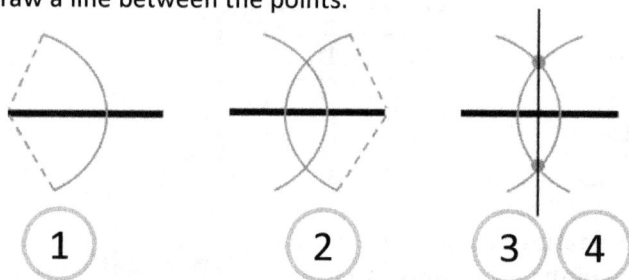

To draw an angle bisector:
1. Place the compass point on the vertex and draw an arc that crosses both sides of the angle.
2. Place the compass point on the intersection of the arc and one of the sides and draw another arc toward the center of the angle.
3. Repeat step 2 on the intersection of the first arc and the other side.
4. Place a point where the second and third arcs intersect.
5. Draw a line from the arc intersection to the vertex.

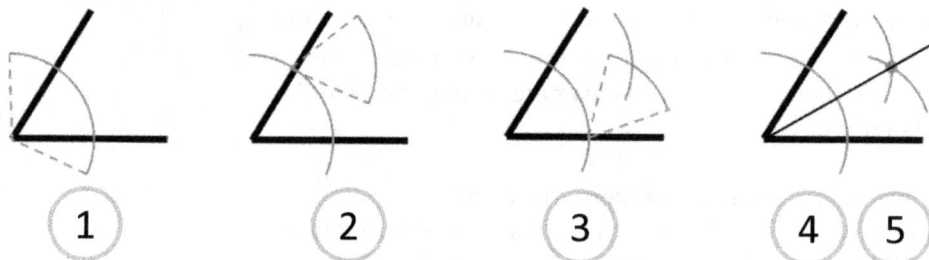

Chapter 2: Theorems & Proofs

Definitions:

Theorem: A statement that has been proven correct.
Corollary: A statement implied or directly derived from a theorem that requires little or no proof.
Postulate or Axiom: A statement that is assumed true without a proof.
Proof: An outline of reasoning that shows a statement is true or false.

Deductive Reasoning: A step by step series of logical arguments to determine if something is true.
> Jane read that entire book.
> That book has 300 pages.
> Therefore, Jane read 300 pages.

Indirect Reasoning: There are two methods -
- Eliminating all possibilities but one which must be true.
> Someone ate a slice of my apple pie.
> Only Joe, Joan, and John had access to it.
> I was talking to Joe when it went missing.
> Joan is allergic to apples.
> Therefore, John must have eaten the slice.

- Proving that the opposite of a statement is not true and concluding that the original statement must then be true.
> Suppose Judy really likes cats.
> I will assume Judy hates cats.
> One day, I see Judy petting a cat and smiling.
> I must conclude Judy does not hate cats.
> Therefore, Judy must like cats.

Proof by Counterexample: Disproving a statement by finding one case in which the statement is untrue.
> All vegetables are red.
> This carrot is a vegetable.
> This carrot is orange.
> Therefore, not all vegetables are red.

Logic Statement: A statement deemed true or false.

If –Then: A statement that says if statement p is true, then statement q is true. Written $p \rightarrow q$ (or p implies q)

If and Only If (iff): A statement that says if p is true, q is true AND vice versa. Written $p \leftrightarrow q$

Logic Operator: An action performed on a statement.

Truth Table: A table that shows the outcome of logic operators on statements.

p	$\sim p$
T	F
F	T

NOT: The logic operator that **negates** a statement.
Notation: $\sim p$ means the opposite of statement p.
p = "The sky is blue."
$\sim p$ = "The sky is not blue."

p	q	$p \wedge q$
T	T	T
T	F	F
F	T	F
F	F	F

AND: The logic operator that combines two statements in a **conjunction**. It asks whether *both* statements are true according to the truth table to the right. Notation: $p \wedge q$ means statement p AND statement q.
p = "The sky is blue."
q = "The ocean is blue."
$p \wedge q$ = "The sky is blue and the ocean is blue."

p	q	$p \vee q$
T	T	T
T	F	T
F	T	T
F	F	F

OR: The logic operator that combines two statements in a **disjunction** asking whether *either* statement is true according to the truth table to the right. Notation: $p \vee q$ means statement p OR statement q.
p = "The sky is blue."
q = "The sky is black."
$p \vee q$ = "The sky is blue or the sky is black."

Truth tables for NOT, AND, and OR.

To use the truth tables, find the correct row for the values of p and q; the last column gives the answer for the operator on those values.

Example

If p is true, and q is false, find $(\sim p \wedge q) \vee p$.

$\sim p = \sim T = F$; $(\sim p \wedge q) = F \wedge F = F$

$(\sim p \wedge q) \vee p = F \vee T = T$

Some Starting Postulates:

Here are some useful postulates that are basically common sense:

> **Two points determine a line.**

> **Three noncollinear points determine a plane.**

> **Two lines can intersect at only one point.**

> **Two planes can intersect at a line.**

> **The length of two non-overlapping line segments is the sum of the length of each line segment.**

> **The total angle measurement of two adjacent angles is the sum of each angle's measurement.**

> **Given a point and a line, there is only one line that can be drawn through the point and perpendicular to the line.**

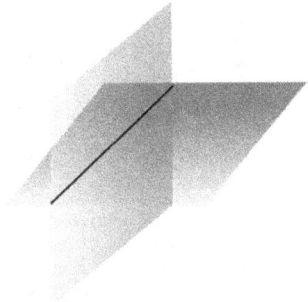

Intersection of two planes.

A $\quad$ B $\quad$ C

$AC = AB + BC$

Addition of two line segments.

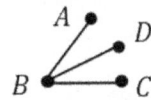

$m\angle ABC =$
$m\angle ABD + m\angle DBC$

Addition of two adjacent angles.

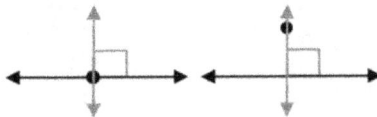

Two cases to illustrate the last postulate.

Example

Let $p = $ "two lines intersect at one point,"
and $q = $ "a line is determined by only one point."
Determine $(p \wedge \sim q) \vee q$.

According to the postulates on the prior page:
$$p = T; \quad q = F$$

Using the truth tables:

$$\sim q = \sim F = T$$

$$(p \wedge \sim q) = T \wedge T = T$$

$$(p \wedge \sim q) \vee q = T \vee F = T$$

In English, this says:

It is true that the statement "a line is determined by only one point" is false $(\sim q = \sim F = T)$.

It is true that two lines intersect at a point AND a line is NOT determined by only one point $((p \wedge \sim q) = T \wedge T = T)$.

It is true that either the above statement OR a line is determined by only one point is true $((p \wedge \sim q) \vee q = T \vee F = T)$.

Chapter 3: Vertical Angles & Parallel Lines

Vertical Angles:

With the foundation laid, we can finally do some Geometry!

If you draw two lines that intersect, the opposing angles, called **vertical angles**, are equal in measure.

> **Vertical angles are congruent.**

We'll prove it with a **two column proof** – a proof that puts assertions and reasoning on the left with justification on the right. To create the proof, we start with what we know (the given)j and state what we need to prove. We usually draw a helpful figure, as well.

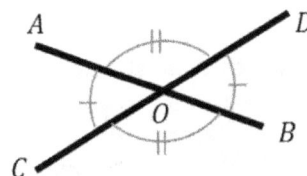

Vertical Angles.

Given: Two lines intersect.
Prove: Vertical angles are congruent.

Statement	Justification
$m\angle AOC + m\angle COB = 180$	Supplementary Angles
$m\angle DOB + m\angle COB = 180$	Supplementary Angles
$m\angle COB = 180 - m\angle DOB$	Algebra
$m\angle AOC + (180 - m\angle DOB) = 180$	Substitution
$m\angle AOC - m\angle DOB = 0$	Algebra
$m\angle AOC = m\angle DOB$	Algebra

You can do the same proof for $m\angle AOD$ and $m\angle COB$.

Parallel Lines:

When two lines are cut by a **transversal** (a line that intersects both lines), several types of angles are formed:

Corresponding Angles
$\angle a$ & $\angle e$; $\angle c$ & $\angle g$; $\angle b$ & $\angle f$; $\angle d$ & $\angle h$

Alternating Interior Angles
$\angle c$ & $\angle f$; $\angle d$ & $\angle e$

Alternating Exterior Angles
$\angle a$ & $\angle h$; $\angle b$ & $\angle g$

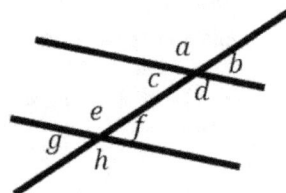

Two parallel lines cut by a transversal.

A postulate states:

> **When two parallel lines are cut by a transversal**
> **Corresponding angles are equal.**

This makes sense since the angle formed by a line and a transversal should be the same as the angle formed by a line parallel to the first and the same transversal.

As long as you have parallel lines, the following pairs of angles are then congruent:

$$\angle a \cong \angle e; \quad \angle c \cong \angle g;$$
$$\angle b \cong \angle f; \quad \angle d \cong \angle h$$

A theorem states:

> **When two parallel lines are cut by a transversal**
> **Alternating interior angles are equal.**

This means

$$\angle c \cong \angle f; \angle d \cong \angle e$$

Let's prove it.

Given: Two parallel lines are cut by a transversal

Prove: $\angle c \cong \angle f$

$\angle d \cong \angle e$

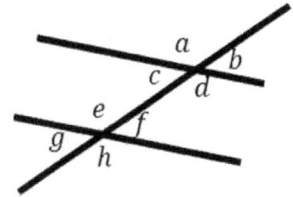

Two parallel lines cut by a transversal.

Statement	Justification
$m\angle a + m\angle c = 180$	Supplementary Angles
$m\angle a = 180 - m\angle c$	Algebra
$m\angle e + m\angle f = 180$	Supplementary Angles
$m\angle e = 180 - m\angle f$	Algebra
$m\angle a = m\angle e$	Corresponding Angles
$180 - m\angle c = 180 - m\angle f$	Substitution (2nd & 4th line)
$m\angle c = m\angle f$	Algebra
$m\angle d + m\angle b = 180$	Supplementary Angles
$m\angle e + m\angle f = 180$	Supplementary Angles
$m\angle b = m\angle f$	Corresponding Angles
$180 - m\angle d = 180 - m\angle e$	Substitution
$m\angle d = m\angle e$	Algebra

Another theorem states:

> **When two parallel lines are cut by a transversal**
> **Alternating exterior angles are equal.**

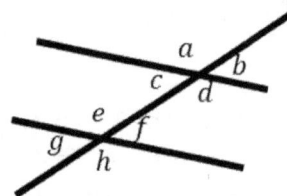

Two parallel lines cut by a transversal.

This means

$$\angle a \cong \angle h; \ \angle b \cong \angle g$$

Let's prove it.

Given: Two parallel lines are cut by a transversal
Prove: $\angle a \cong \angle h$
$\angle b \cong \angle g$

Statement	Justification
$m\angle a = m\angle e$	Corresponding Angles
$m\angle h = m\angle e$	Vertical Angles
$m\angle a = m\angle h$	Transitive Property
$m\angle b = m\angle f$	Corresponding Angles
$m\angle g = m\angle f$	Vertical Angles
$m\angle b = m\angle g$	Transitive Property

With these handy relationships, we can figure out a lot
from a figure. Let's look at an example.

Example

If $\angle a = 150°$, find all of the other angles.

We know corresponding angles are congruent:

$\angle a \cong \angle e = 150°$;

We know that vertical angles are congruent:

$\angle a \cong \angle d = 150°$;

$\angle e \cong \angle h = 150°$;

We know adjacent angles add, and that straight

angles are $180°$:

$\angle a + \angle b = 180°$;

$\angle b = 180° - \angle a = 30°$;

We know corresponding angles are congruent:

$\angle b \cong \angle f = 30°$;

We know that vertical angles are congruent:

$\angle b \cong \angle c = 30°$;

$\angle f \cong \angle g = 30°$;

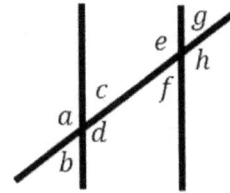

Two parallel lines cut by a transversal.

Proving Parallelism:

It is probably not too surprising that the way to prove two lines are parallel is to cut them with a transversal and look at the relationship of the angles. We have an important postulate that states:

Two lines are parallel if when cut by a transversal, corresponding angles are congruent.

Based on that postulate, we can put forth another set of theorems which we'll summarize as follows:

> **Two lines are parallel if when cut by a transversal one of the following can be proved:**
> - **A pair of alternate interior angles are congruent;**
> - **A pair of alternate exterior angles are congruent;**
> - **A pair of same side interior angles are supplementary;**
> - **A pair of same side exterior angles are supplementary.**

Proofs of the above statements are left as problems.

Example

Are the two vertical lines shown parallel?

By inspection, we see that $\angle a$ is supplementary to the angle measuring 150°. Therefore

$$\angle a = 180° - 150° = 30°.$$

As a result, we have a pair of congruent corresponding angles, so by the postulate, the vertical lines are parallel.

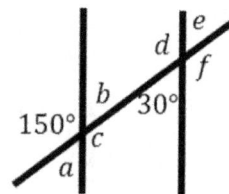

Figure for example.

Chapter 4: Triangles

Definitions:

Triangle: A three-sided, closed figure. Notation: $\triangle ABC$.
Vertex: The points of the triangle .
Median: The line connecting one vertex of a triangle with
  the midpoint of the opposite side.
Altitude: The line from one vertex of a triangle that
  intersects the opposite side (or extension of that side)
  at a 90° angle.

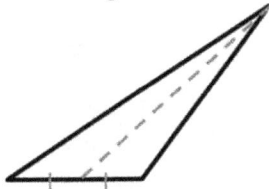
Median.

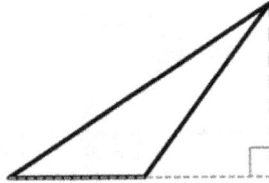
Altitude.

Acute Triangle: A triangle whose largest angle is
  less than 90°.
Right Triangle: A triangle whose largest angle is
  equal to 90°.
Obtuse Triangle: A triangle whose largest angle is
  more than 90°.

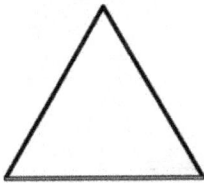
Acute.

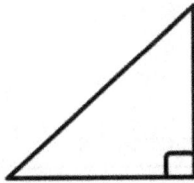
Right.

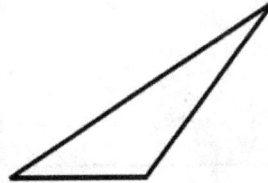
Obtuse.

Equilateral Triangle: A triangle with three equal sides.
Isosceles Triangle: A triangle with two equal sides.
Scalene Triangle: A triangle with no equal sides.

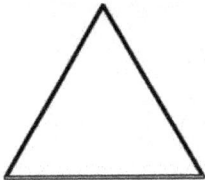
Equilateral.

Isosceles.

Scalene.

17

Properties of Triangles:

A very useful theorem about **interior angles** (angles inside the triangle perimeter) states:

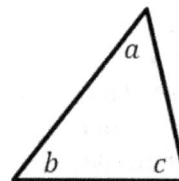

Triangle.

> **The interior angles of a triangle add to 180°.**

Before we prove this, let's do some manipulation to the figure of the triangle. We will draw a line through vertex a that is parallel to the base. We do this to create two systems of parallel lines cut by a transversal:

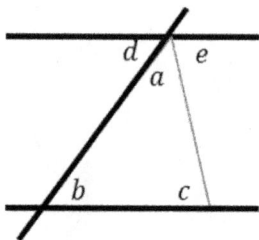

System 1.

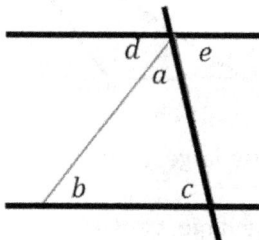

System 2.

Given: There is a triangle
Prove: $\angle a + \angle b + \angle c = 180°$

Statement	Justification
$m\angle d + m\angle a + m\angle e = 180$	Supplementary Angles
$m\angle b = m\angle d$	Alt. Interior Angles
$m\angle c = m\angle e$	Alt. Interior Angles
$m\angle b + m\angle a + m\angle c = 180$	Substitution

Several corollaries can be drawn from the triangle angle addition theorem:

> **A triangle can have no more than one angle that is greater than or equal to 90°.**

> **For a right triangle, the other two angles are complementary.**

> **In comparing two triangles, if two of their angles are congruent, then their third angles are congruent.**

We'll leave the proofs as problem exercises. We can make another observation about the triangle angles, this time about the **exterior angles** (the supplements of interior angles):

> **An exterior angle of one vertex of a triangle is equal to the sum of the other two interior angles.**

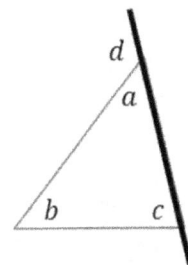

Triangle with exterior angle $\angle d$.

Given: There is a triangle
Prove: $m\angle d = m\angle b + m\angle c$

Statement	Justification
$m\angle d + m\angle a = 180$	Supplementary Angles
$m\angle b + m\angle c + m\angle a = 180$	Sum Interior Angles
$m\angle a = 180 - m\angle b - m\angle c$	Algebra
$m\angle d + (180 - m\angle b - m\angle c) = 180$	Substitution
$m\angle d = m\angle b + m\angle c$	Algebra

Triangle Inequalities:

From the preceding theorem, we can see that if an exterior angle is the sum of the other two angles of triangle, it should be larger than either of those two angles alone:

> **The exterior angle of a vertex of a triangle is larger than either of the other two interior angles.**

There is a similar conclusion we can make about the sides of a triangle:

> **Any side in a triangle is smaller than the sum of the other two sides.**

We can intuit this from the following argument- If the shortest distance between two points is a straight line, then:

$$\overline{AC} < \overline{AB} + \overline{BC}.$$

Finally, we state a set of theorems whose proofs will be given in Appendix D:

> **If two sides of a triangle are not congruent:**
> - **The opposing angles are not congruent, and**
> - **The larger angle is opposite the larger side.**

> **If two angles of a triangle are not congruent:**
> - **The opposing sides are not congruent, and**
> - **The larger side is opposite the larger angle.**

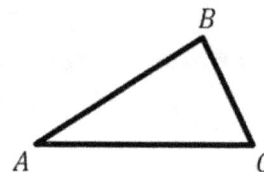

Triangle Inequalities:

$$\overline{AC} < \overline{AB} + \overline{BC}$$

$$\overline{BC} < \overline{AC} \leftrightarrow \angle A < \angle B.$$

Another way to state the above is:

$$\overline{BC} < \overline{AC} \leftrightarrow \angle A < \angle B.$$

For the figure given, which is larger, side $\overline{AB}$ or $\overline{AC}$?

By the sum of angles of triangles, we know that:

$$\angle A + \angle B + \angle C = 180°$$

We also know from the figure that $\angle A = 90°$ and $\angle C = 30°$. Therefore:

$$90° + \angle B + 30° = 180°$$

$$\angle B = 60°$$

Since $\angle B > \angle C$, we must conclude that $\overline{AC} > \overline{AB}$.

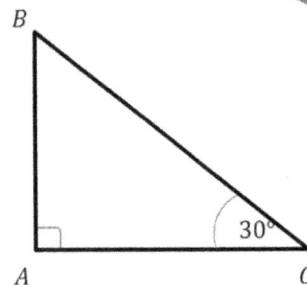

Triangle for example.

For the figure given, find $m\angle BAC$.

By supplementary angles:

$$\angle BCA + \angle ACD = 180°$$
$$\angle ACD = 180° - 120° = 60°$$

By triangle angle sum:

$$\angle DAC + \angle ACD + \angle D = 180°$$
$$\angle DAC = 180° - 60° - 90° = 30°$$

By addition of angles:

$$\angle EAB + \angle BAC + \angle CAD = 90°$$
$$\angle BAC = 90° - 40° - 30° = 20°$$

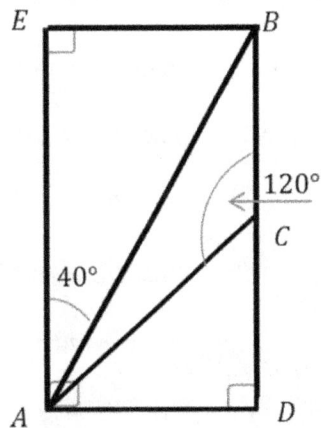

Triangles for example.
$\angle EAD = 90°$.

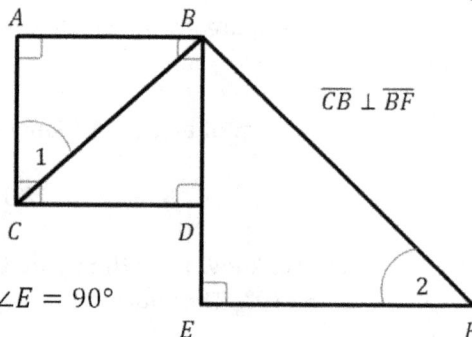

Figure for example.

For the figure given, prove $\angle 1 \cong \angle 2$.

$\overline{CB} \perp \overline{BF}$

Given: $\overline{CB} \perp \overline{BF}$

$\angle A \cong \angle D \cong \angle ACD \cong \angle ABD \cong \angle E = 90°$

Prove: $\angle 1 \cong \angle 2$

Statement	Justification
$m\angle 1 + m\angle A + m\angle ABC = 180$	Triangle Angle Sum
$m\angle ABC = 90 - m\angle 1$	$\angle A = 90°$ (Given)
$m\angle CBD = 90 - m\angle ABC$	Complimentary Angles
$m\angle CBD = 90 - (90 - m\angle 1)$	Substitution
$m\angle CBD = m\angle 1$	Algebra
$m\angle CBD + m\angle EBF = 90$	Perpendicular Lines
$m\angle 1 + m\angle EBF = 90$	Substitution
$m\angle EBF = 90 - m\angle 1$	Algebra
$m\angle EBF + m\angle 2 + m\angle E = 180$	Triangle Angle Sum
$m\angle EBF + m\angle 2 = 180 - 90 = 90$	$\angle E = 90°$ (Given)
$90 - m\angle 1 + m\angle 2 = 90$	Substitution
$\angle 1 \cong \angle 2$	Algebra

Chapter 5: Congruent Triangles

Congruent Triangles:

Triangles are made up of three sides and three angles that must form a closed figure. As a result, triangles obey a **Rigidity Property**: if a certain set of three quantities are given, there is only one way to construct the triangle, and therefore the other three quantities can be deduced. The sets of given quantities that meet this criteria include:

> **Side Side Side (SSS)** –
> > knowing all three sides
>
> **Side Angle Side (SAS)** –
> > knowing two sides and the angle between them
>
> **Angle Side Angle (ASA)** –
> > knowing two angles and the side between them
>
> **Angle Angle Side (AAS)** –
> > knowing two angles and one adjacent side

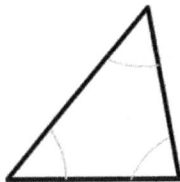
SSS

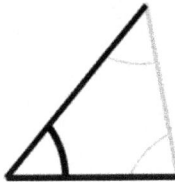
SAS

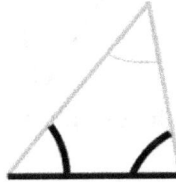
ASA

AAS

The remaining combinations are not valid. To remember this: Angle Angle Angle is an auto-club, and Angle Side Side has an unfortunate abbreviation.

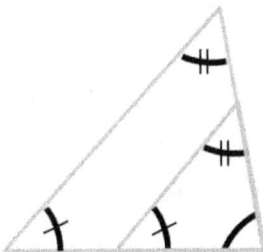
AAA can produce multiple triangles.

SSA can produce two triangles.

> ### Special Exception:
>
> SSA fails because one can draw an acute or obtuse triangle from the known quantities. For a right triangle only, SSA is a valid option.

Knowing all of this, we can now determine if two different triangles are **congruent**, i.e. if the sides and angles of one triangle are equal to the sides and angles of another triangle.

If there is only one way to draw a triangle given three known quantities, then if two triangles share those three quantities, their other three quantities must also be equal.

Again, we can employ **SSS**, **SAS**, **ASA**, and **AAS**.

> **Example** For each triangle below, determine if it is congruent to the triangle to the right, and give justification.

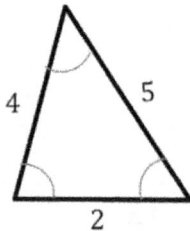

Comparison Triangle.

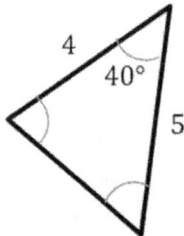

Congruent: SSS

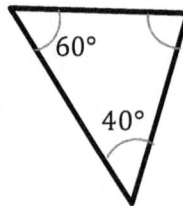

Indeterminate: AAA

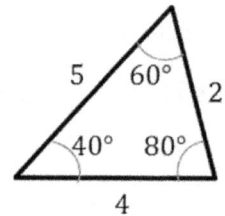

Congruent: SAS

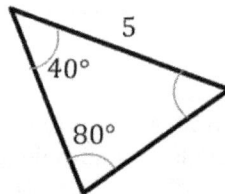

Congruent: AAS

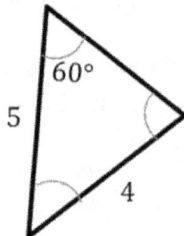

Indeterminate: SSA

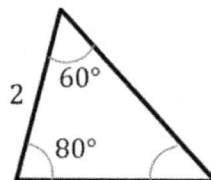

Congruent: ASA

Isosceles Triangles & Bisectors:

We can use triangle congruency to develop some interesting properties about bisectors and the isosceles triangle.

> **Any point on a perpendicular bisector of a line segment is equidistant from the endpoints of the line segment.**

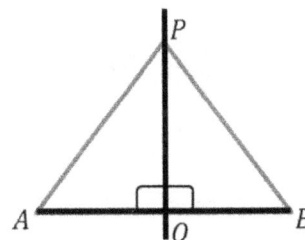

Perpendicular bisector.

Given: $\overline{PO}$ is a perpendicular bisector of $\overline{AB}$
 P is a point on the perpendicular bisector
Prove: $\overline{AP} \cong \overline{PB}$

Statement	Justification
$m\angle AOP = m\angle POB = 90$	*Perpendicular* Bisector
$\overline{AO} \cong \overline{OB}$	Perpendicular *Bisector*
$\overline{OP} \cong \overline{OP}$	Reflexive Property
$\triangle AOP \cong \triangle BOP$	SAS
$\overline{AP} \cong \overline{PB}$	Congruent Triangles

Notice that $\triangle APB$ is an isosceles triangle since $\overline{AP} \cong \overline{PB}$. As it is made up of two congruent triangles, we can also state that $\angle A \cong \angle B$. Therefore:

> **Two sides of a triangle are congruent iff their opposing angles are congruent.**

Note the iff (if and only if) statement means if the sides are equal, the angles are equal, and vice versa.

Another theorem about bisectors states:

> **Any point on an angle bisector is equidistant from the sides of the angle.**

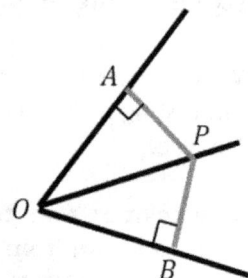

Angle bisector.

Given: $\overline{PO}$ is a bisector of $\angle AOB$
  P is a point on the bisector
  $\overline{AP} \perp \overline{OA}; \overline{BP} \perp \overline{OB}$ (marks distance from P to angle sides)
Prove: $\overline{AP} \cong \overline{PB}$

Statement	Justification
$m\angle PAO = m\angle PBO = 90$	Perpendicular Lines
$m\angle AOP = m\angle POB$	Angle Bisector
$\overline{OP} \cong \overline{OP}$	Reflexive Property
$\triangle AOP \cong \triangle POB$	AAS
$\overline{AP} \cong \overline{PB}$	Congruent Triangles

Example

Given the figure to the right, and that
$\overline{AC} \cong \overline{CE}; \overline{AG} \cong \overline{FE}; \angle HGF \cong \angle HFG$;
show that $\overline{GD} \cong \overline{FB}$.

Because of the isosceles angle/side theorem,
$\overline{AC} \cong \overline{CE}$ tells us that $\triangle ACE$ is isosceles and
therefore $\angle A \cong \angle E$.

Figure for example.

Since $\overline{AG} \cong \overline{FE}$, $\overline{GF} \cong \overline{GF}$, and
$\overline{AF} = \overline{AG} + \overline{GF}$ and $\overline{GE} = \overline{GF} + \overline{FE}$, then
$\overline{AF} \cong \overline{GE}$.

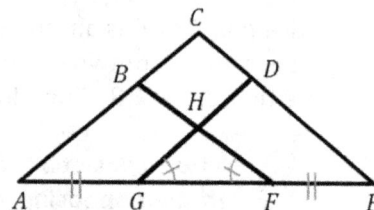

Finally, with $\angle HGF \cong \angle HFG$, we have
$\triangle ABF \cong \triangle GDE$ by ASA, and therefore $\overline{GD} \cong \overline{FB}$.

Chapter 6: Similar Triangles

Similar Triangles:

A triangle can be a scaled version of another triangle. The angles of both triangles are the same, but their sides differ by a multiplicative factor. In this case, we say the triangles are **similar**:

$$\Delta ABC \sim \Delta DEF$$

If we know two of the angles of a triangle, we must know the third (all angles add to $180°$). This means that if two triangles have two congruent angles, then their third must be congruent. This in turn means the triangles are similar. We have employed the Angle Angle Angle (AAA) relationship to prove **similitude**. This can be summarized as follows:

> **AA - If two triangles have two corresponding congruent angles, the triangles are similar.**

Again, we only need to know two of the angles are congruent because of the sum of interior angles theorem.

There are two other ways to determine similitude:

> **SSS – If all corresponding sides have the same <u>proportion</u>, then the triangles are similar**
>
> **SAS – If two sides have the same <u>proportion</u>, and the included corresponding angles are congruent, then the triangles are similar.**

The first theorem stems from how we defined similitude; the other stems from the rigidity property of triangles.

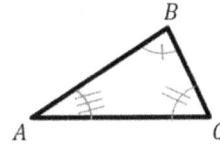

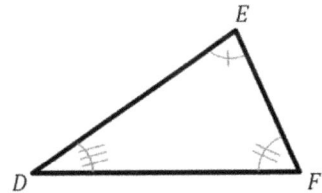

For each triangle below, determine if it is similar to the triangle to the right, and give justification.

5" 60°
2"
40° 80°
4"
Comparison triangle.

60°

40°

This triangle is similar by AA.

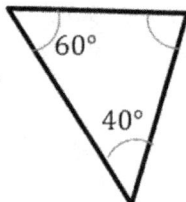

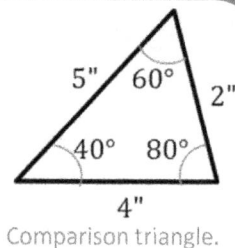

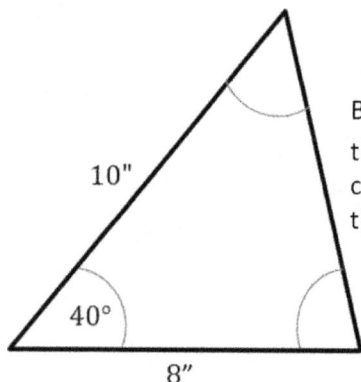

10"

40°

8"

Because $\frac{10}{5} = \frac{8}{4}$, and because the angles between the corresponding sides match, this triangle is similar by SAS.

10"

7"

Because $\frac{10}{5} = \frac{8}{4} \neq \frac{7}{2}$, this triangle is NOT similar by SSS.

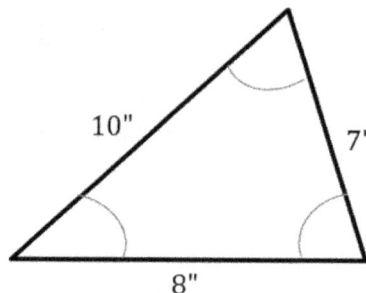

8"

Properties of Similar Triangles:

Now let's draw a line through a triangle so that it is parallel to one of the sides. We can draw a couple of conclusions:

> **Drawing a line through a triangle parallel to one of its sides produces a similar triangle to the first.**

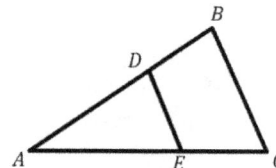

B
D
A   E   C
Line drawn through triangle parallel to one side.

Given: $\overline{DE} \parallel \overline{BC}$
Prove: $\triangle ABC \sim \triangle ADE$

Statement	Justification
$\angle A \cong \angle A$	Reflexive Property
$\angle ABC \cong \angle ADE$	Corresponding Angles of parallel lines cut by transversal
$\triangle ABC \sim \triangle ADE$	AA Similitude

And, because the two triangles are similar, we can say that side $\overline{DE}$ is proportional to side $\overline{BC}$ written:

$$\overline{DE} \propto \overline{BC}$$

In fact, because of similitude, all sides share the same proportion or ratio. This can be written as:

$$\frac{AD}{AB} = \frac{AE}{AC} = \frac{DE}{BC} = x$$

The **perimeter** of any shape is the sum of its sides. Looking at $\triangle ABC$ and $\triangle ADE$, and using the above ratio, we find:

> The **perimeter** of a $\triangle$ is the sum of its sides.

$$AB + BC + AC = Perim\ \triangle ABC$$

$$AD + DE + AE = xAB + xBC + xAC = Perim\ \triangle ADE$$

$$x(AB + BC + AC) = x(Perim\ \triangle ABC) = Perim\ \triangle ADE$$

$$\frac{Perimeter\ of\ \triangle ADE}{Perimeter\ of\ \triangle ABC} = x$$

The perimeters of two similar triangles are in the same proportion as their sides.

Finally, let's draw an altitude on two similar triangles. A theorem states:

> **The altitudes of two similar triangles are in the same proportion as their sides.**

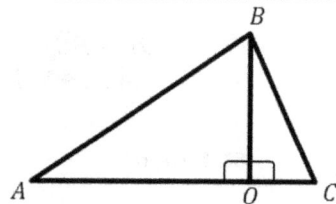

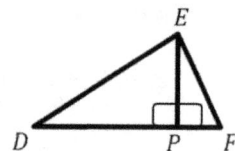

Altitudes drawn for two similar triangles.

Given: $\triangle ABC \sim \triangle DEF$
$\overline{BO} \perp \overline{AC}$; $\overline{EP} \perp \overline{DF}$
Prove: $\dfrac{AB}{DE} = \dfrac{BO}{EP}$

Statement	Justification
$\angle A \cong \angle D$	Similar Triangles
$\angle BOA \cong \angle EPD$	Both are 90°
$\triangle AOB \sim \triangle DPE$	AA Similitude
$\dfrac{AB}{DE} = \dfrac{BO}{EP}$	Similar Triangles

Finally, the **area** of a triangle is ½ times the **base** (a side) times altitude. We know from above that:

$$\frac{AB}{DE} = \frac{AC}{DF} = \frac{BO}{EP} = x$$

> **The area** of a △ is
> $$area = \frac{bh}{2},$$
> b = base;
> h = height or altitude.

Plugging into area, we get:

$$\frac{1}{2}(AC)(BO) = Area\ \triangle ABC$$
$$\frac{1}{2}(DF)(EP) = \frac{1}{2}(xAC)(xBO) = Area\ \triangle DEF$$

$$\frac{Area\ of\ \triangle ABC}{Area of\ \triangle DEF} = x^2$$

> **The areas of two similar triangles are in proportion to the square of the proportion of their sides.**

Example

Given that the altitude (vertical direction) of triangle A is 3.2", determine the third side, altitude, perimeter, and area of triangle B.

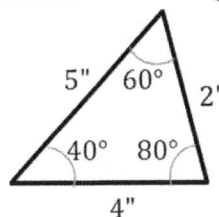

Triangle A.

First, we make sure the triangles are similar:

Comparing sides, $\frac{10}{5} = \frac{8}{4} = 2$, and the angles between the corresponding sides match, therefore the triangles are similar by SAS.

The sides and altitude are in equal proportion, so the third side of triangle B is $2 \times 2" = 4"$, and the altitude is $2 \times 3.2" = 6.4"$.

The perimeter of triangle A is $5" + 4" + 2" = 11"$, so the perimeter of triangle B is $2 \times 11" = 22"$. (Check: $10 + 8 + 4 = 22$.)

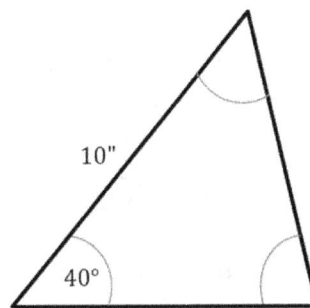

Triangle B.

The area of triangle A is $\frac{4 \times 3.2}{2} = 6.4$ in², so the area of triangle B must be $(2^2) \times 6.4 = 25.6$ in². (Check: $\frac{8 \times 6.4}{2} = 25.6$.)

Triangle Midsegments:

We can use similar triangles to prove some relationships about triangle **midsegments** – lines that connect the midpoint of one side to the midpoint of another. A theorem states:

> **The midsegment of a triangle drawn between two sides is parallel to the third side and half its length.**

Let's prove it:

Given: $\overline{AO} \cong \overline{OB}$; $\overline{AP} \cong \overline{PC}$
Prove: $\overline{OP} \parallel \overline{BC}$; $BC = 2(OP)$

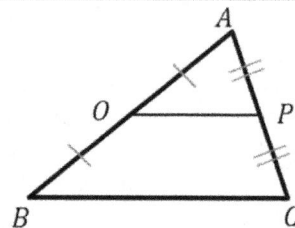

Triangle with midsegment.

Statement	Justification
$\overline{AO} \cong \overline{OB}$; $\overline{AP} \cong \overline{PC}$	Given
$AB = 2AO$; $AC = 2AP$	Segment Addition
$\angle A \cong \angle A$	Reflexive
$\triangle ABC \sim \triangle AOP$	SAS Similitude
$BC = 2OP$	Similar Triangles
$\angle B \cong \angle AOP$; $\angle C \cong \angle OPA$	Similar Triangles
$\overline{OP} \parallel \overline{BC}$	Corresp. Angles of Parallel lines & transversal

If we draw all three midsegments (each midsegment is half the corresponding side), then we have created a similar triangle within the first called a **midsegment triangle**. Because of similitude, we can say:

> **The perimeter of a midsegment triangle is half the perimeter of the original triangle.**

> **The area of a midsegment triangle is one quarter the area of the original triangle.**

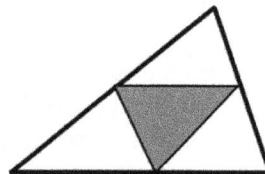

Triangle with midsegment triangle inside.

The proofs will be left as exercises.

Chapter 7: Right Triangles

Right Triangles:

A right triangle is characterized by a right angle at one of its vertices. The leg across from the right angle is called the **hypotenuse**. A theorem states:

> **If an altitude is drawn to the hypotenuse of a right triangle, two similar right triangles are formed which are also similar to the first.**

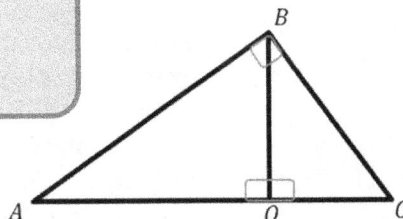

Given: $\triangle ABC$ is a right triangle
$\overline{BO} \perp \overline{AC}$
Prove: $\triangle AOB \sim \triangle BOC \sim \triangle ABC$

Altitude drawn to hypotenuse
of a right triangle.

Statement	Justification
$m\angle BOA = m\angle BOC = 90°$	Perpendicular Lines
$m\angle A + m\angle ABO = 90°$	Complementary Angles of Right $\triangle$
$m\angle C + m\angle A = 90°$	
$(90° - m\angle C) + m\angle ABO = 90°$	Substitution
$m\angle C = m\angle ABO$	Algebra
$\triangle AOB \sim \triangle BOC$	AA Similitude (1st & 5th lines)
$m\angle BOA = m\angle ABC = 90°$	Right Angles
$m\angle A = m\angle A;\ m\angle C = m\angle C$	Reflexive Property
$\triangle AOB \sim \triangle BOC \sim \triangle ABC$	AA Similitude

Example

Determine x in the figure to the right.

From the figure, we see that $\overline{OB}$ is an altitude of a right triangle, therefore:

$$\triangle ABC \sim \triangle AOB \sim \triangle BOC.$$

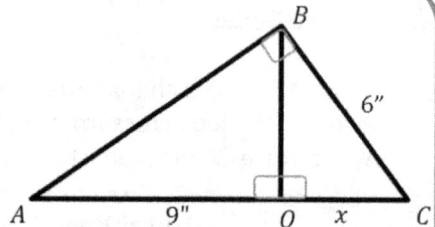

Figure for example.

To make things less confusing, it is helpful to mark equal angles on the diagram:

In $\triangle ABC$, $m\angle C + m\angle A = 90°$
In $\triangle AOB$, $m\angle ABO + m\angle A = 90°$
In $\triangle BOC$, $m\angle CBO + m\angle C = 90°$

Therefore:
$$m\angle A = m\angle CBO$$
$$m\angle C = m\angle ABO.$$

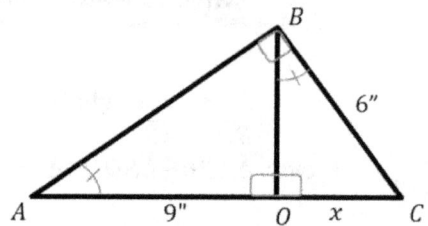

Figure with two congruent angles marked.

With the first set of congruent angles marked, it is easier to align the ratios of similar sides. We'll look at the two triangles that contain the sides of interest (those with known values or the quantity we need):

$$\triangle ABC \sim \triangle BOC \;\to\; \frac{AB}{BO} = \frac{BC}{OC} = \frac{AC}{BC}$$

$$\frac{AB}{BO} = \frac{6}{x} = \frac{9+x}{6}$$

Solving the last two terms for x we get:

$$x^2 + 9x - 36 = 0$$
$$(x-3)(x+12) = 0$$
$$x = \{3, -12\}$$

The only answer that makes physical sense (i.e. has a positive length) is $x = 3$".

Knowing these triangles are similar and aligning their congruent angles, we can write the following relationships (for ease of following along the three triangles are broken apart to the right):

$$\frac{c_1}{h} = \frac{a}{b} = \frac{h}{c_2} \text{ from } \Delta AOB \sim \Delta BOC$$

$$\frac{c_1}{a} = \frac{h}{b} = \frac{a}{c} \text{ from } \Delta AOB \sim \Delta ABC$$

$$\frac{h}{a} = \frac{c_2}{b} = \frac{b}{c} \text{ from } \Delta BOC \sim \Delta ABC$$

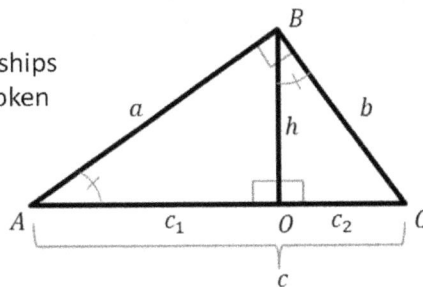

Altitude drawn to hypotenuse of a right triangle.

With these equations, we can derive a relationship between the sides of a right triangle:

$$c_1 = \frac{a^2}{c} \qquad \text{from the 2}^\text{nd} \text{ set of equations}$$

$$c_2 = \frac{b^2}{c} \qquad \text{from the 3}^\text{rd} \text{ set of equations}$$

$$c_1 + c_2 = \frac{1}{c}(a^2 + b^2) \text{ adding the above equations}$$

$$c = \frac{1}{c}(a^2 + b^2) \qquad \text{noting that } c = c_1 + c_2$$

$$c^2 = a^2 + b^2 \qquad \text{simplifying}$$

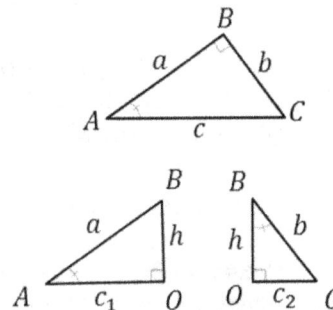

The three triangles separated.

The final result is **Pythagorean's Theorem**:

The square of the hypotenuse of a right triangle is equal to the sum of the squares of the other two sides.

$$c^2 = a^2 + b^2$$

The values $\{a, b, c\}$ that obey Pythagorean's theorem are called **Pythagorean Triples**.

Example

Find the third side of the triangle to the right.

The triangle is a right triangle, so it must obey Pythagorean's Theorem:

$$5^2 = 4^2 + x^2$$

$$x = \sqrt{25 - 16} = \sqrt{9} = 3"$$

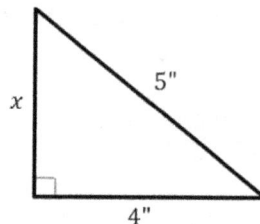

Right triangle for example.

Special Right Triangles:

Common Pythagorean Triples: A few Pythagorean triples are sets of whole number values, and are therefore used quite often. The three most common ones are:

$\{3,4,5\}$, $(5,12,13)$, and $\{8,15,17\}$

The 45-45-90 Triangle: A right triangle whose other two angles are each 45° have a special relationship among the sides. Because the two other angles are equal, so are their opposing sides – they form an isosceles triangle. If $a = b$, then:

$$c^2 = 2a^2 \rightarrow c = a\sqrt{2}$$

The ratio of sides is then $a : b : c = 1 : 1 : \sqrt{2}$.

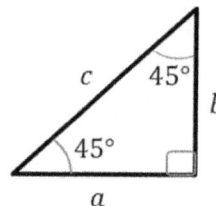

The 45-45-90 triangle.

The 30-60-90 Triangle: The 30-60-90 right triangle also has a special relationship between sides. To determine this, we place two identical triangles back to back. Since the adjacent angles add to 60°, we have an equilateral triangle (all sides of the large triangle are equal). So, $c = 2a$, and:

$$c^2 = 4a^2 = a^2 + b^2 \rightarrow b = a\sqrt{3}.$$

The ratio of sides is then $a : b : c = 1 : \sqrt{3} : 2$.

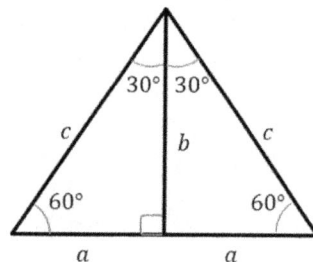

Two 30-60-90 triangles.

Example

Find the third side of the triangle to the right.

Recognizing that the two sides are equal, this is a 45-45-90 triangle: $a : b : c = 1 : 1 : \sqrt{2}$.

Since $a = b = 3$", then $c = x = 3\sqrt{2}$".

Check: $(3\sqrt{2})^2 = 18$
$$3^2 + 3^2 = 18$$

Pythagorean's Theorem holds.

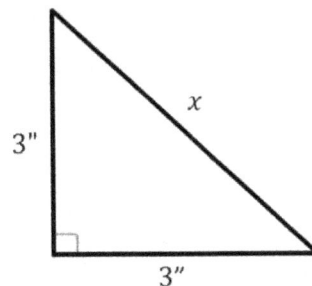

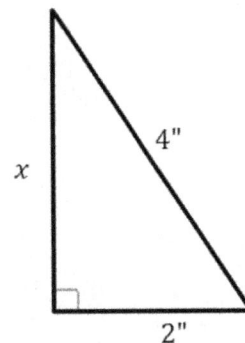

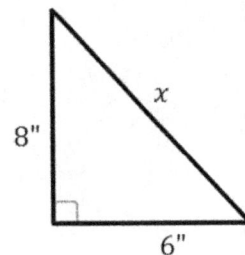

Right triangle for example.

Example

Find the third side of the triangle to the right.

Recognizing that the hypotenuse is twice the side, this is a 30-60-90 triangle: $a : b : c = 1 : \sqrt{3} : 2$.

Since $a = 2$", $c = 4$", then $b = x = 2\sqrt{3}$".

Check: $4^2 = 16$
$$2^2 + (2\sqrt{3})^2 = 4 + 12 = 16$$

Pythagorean's Theorem holds.

Right triangle for example.

Example

Find the third side of the triangle to the right.

Recognizing that the two sides are similar to a 3:4:5 triple with a factor of 2:

$$\frac{6}{3} = \frac{8}{4} = \frac{x}{5} \rightarrow x = 10.$$

Check: $(10)^2 = 100$
$$6^2 + 8^2 = 36 + 64 = 100$$

Pythagorean's Theorem holds.

Right triangle for example.

Chapter 8: Polygons

Definitions:

Polygon: A closed figure with multiple vertices and sides.

Name	# Sides	Name	# Sides
Triangle	3	Heptagon	7
Quadrilateral	4	Octagon	8
Pentagram	5	Nonagon	9
Hexagon	6	Decagon	10

Convex Polygon: A polygon which has the property that a line can only cross it at no more than two points.

Concave Polygon: A polygon which has the property that a line can cross it at more than two points.

Regular Polygon: A polygon whose sides are equal and whose angles are equal.

Vertex: Point where line segments meet to form an angle.

Diagonal: A line that connects non-consecutive vertices.

Apothem: The perpendicular from center to one side.

> **Perimeter:** The measure of the sum of all sides.
> **Area:** The measure of the interior. For regular polygons, the $area = (ap/2)$, a =apothem, p =perimeter.

Parallelogram: A quadrilateral whose opposing sides are parallel. The **area** is the length of one side times the perpendicular distance to the opposite side.

Square: A regular quadrilateral.

Rectangle: A parallelogram with four right angles.

Rhombus: A parallelogram with four equal sides.

Trapezoid: A quadrilateral with a set of opposing parallel sides (called **bases**), and a set of opposing nonparallel sides (called **legs**).

Isosceles Trapezoid: A trapezoid with equal legs.

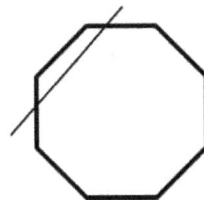

Convex octagon.

Concave octagon.

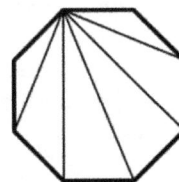

Octagon with five diagonals drawn.

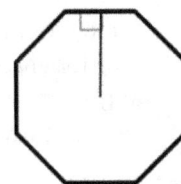

Octagon with apothem drawn.

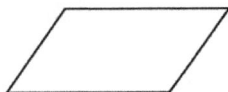

Parallelogram.   Square.   Rectangle.   Rhombus.   Isosceles trapezoid.

Polygon Angle Properties:

Convex polygons obey a rule for the sum of their interior angles:

> **The sum of interior angles of a convex polygon is two less the number of sides multiplied by 180°.**
>
> $$S = (n - 2)180°$$

Interior angles of an octagon indicated.

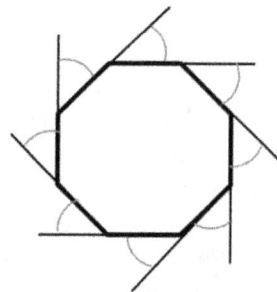

So, the interior angles of a triangle sum to $(3 - 2)180° = 180°$, a quadrilateral sums to $(4 - 2)180° = 360°$, and an octagon sums to $(8 - 2)180° = 1080°$.

Similarly, the exterior angles also have to follow a summation rule:

> **The sum of exterior angles of a convex polygon with one exterior angle at each vertex is 360°.**

Exterior angles of an octagon indicated.

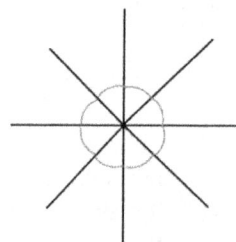

By extension, if a convex polygon is a regular polygon (all angles equal), then each exterior angle has a measure of $360°/n$, with n being the number of sides.

It may seem counter intuitive that an octagon whose interior angles add to 1080° would have the same exterior angle sum as a triangle. But, the exterior angles of a closed shape basically trace out a circle. And, a circle has 360°.

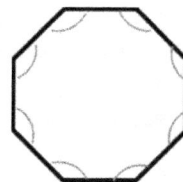

Exterior angles of an octagon rearranged.

Example

What are the interior and exterior angles of a regular nonagon?

A nonagon has nine sides, so the interior angles add to $(9 - 2)180° = 1260°$. Since the nonagon is regular, the interior angles are $\frac{1260°}{9} = 140°$. The exterior angles measure $\frac{360°}{9} = 40°$.

Parallelogram Properties:

We start with special angle properties for a parallelogram:

> **For a parallelogram,**
> **Opposing angles are congruent;**
> **Adjacent angles are supplementary.**

Given: The figure to the right is a parallelogram
Prove: $\angle 2 \cong \angle 3$; $\angle 1 \cong \angle 4$;
$\qquad \angle 1 + \angle 2 = 180°$; $\angle 2 + \angle 4 = 180°$

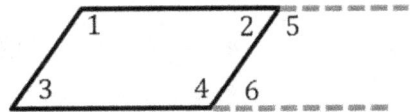

Parallelogram for proof.

Statement	Justification
$\angle 2 \cong \angle 6$; $\angle 4 \cong \angle 5$	Alt. Int. Angles of Parallel Lines & Transversal
$\angle 3 \cong \angle 6$; $\angle 1 \cong \angle 5$	Corresp. Angles of Parallel Lines & Transversal
$\angle 2 \cong \angle 3$	Transitive
$\angle 1 \cong \angle 4$;	Transitive
$\angle 2 + \angle 5 = 180°$	Straight Angle
$\angle 2 + \angle 1 = 180°$	Substitution
$\angle 2 + \angle 4 = 180°$	Substitution

The sides and diagonals of a parallelogram also have special relationships:

> **For a parallelogram,**
> **Opposing sides are congruent;**
> **Diagonals bisect each other.**

Given: The figure to the right is a parallelogram
Prove: $\overline{AC} \cong \overline{BD}$; $\overline{AB} \cong \overline{CD}$;
  $\overline{AO} \cong \overline{OD}$; $\overline{CO} \cong \overline{OB}$;

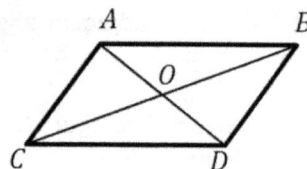

Parallelogram for proof.

Statement	Justification
$\angle CAD \cong \angle ADB$	Alt. Int. Angles of Parallel Lines & Transversal
$\angle ACD \cong \angle ABD$	Opp. Angles of Parallelogram
$\overline{AD} \cong \overline{AD}$	Reflexive
$\triangle ACD \cong \triangle ABD$	AAS Congruency
$\overline{AC} \cong \overline{BD}$; $\overline{AB} \cong \overline{CD}$	Triangle Congruency
$\angle AOB \cong \angle COD$	Vertical Angles
$\angle OAB \cong \angle ODC$	Alt. Int. Angles of Parallel Lines & Transversal
$\triangle OAB \cong \triangle ODC$	AAS with $\overline{AB} \cong \overline{CD}$
$\overline{AO} \cong \overline{OD}$; $\overline{CO} \cong \overline{OB}$	Triangle Congruency

Example

Show that the diagonals of a rectangle are congruent.

We know that because a rectangle is a parallelogram that $\overline{AB} \cong \overline{CD}$ and $\overline{AC} \cong \overline{BD}$. Also, because we are dealing with four right angles, $\angle CAB \cong \angle ABD \cong \angle BDC \cong \angle DCA$.

By SAS, we have that $\triangle ABC \cong \triangle BAD$. This means that $\overline{AD} \cong \overline{BC}$.

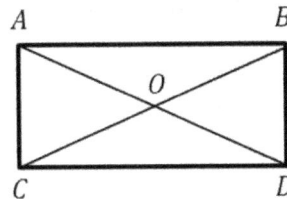

Rectangle for example.

Example

Show that the median to the hypotenuse of a right triangle is half the length of the hypotenuse.

Extending the right triangle to form a rectangle and using the results from the previous example, $\overline{AD} \cong \overline{BC}$. We also know that the diagonals of a parallelogram bisect each other so that $OC = BC/2$.

Using substitution: $OC = AD/2$.

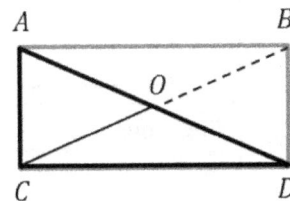

Triangle for example.

Example

Show that the diagonals of a rhombus intersect each other at right angles.

By definition of a rhombus, $\overline{AB} \cong \overline{BD} \cong \overline{CD} \cong \overline{AC}$. Also, because it is a parallelogram, $\overline{AO} \cong \overline{DO}$ and $\overline{CO} \cong \overline{BO}$.

By SSS, $\triangle AOB \cong \triangle AOC$. Because of congruence, $\angle AOB \cong \angle AOC$.

We know that vertical angles are equal so $\angle AOB \cong \angle COD$ and $\angle AOC \cong \angle BOD$. Using substitution, we have that all of the angles at point O are congruent. There are $360°$ around point O, so each angle is

$$\frac{360°}{4} = 90°.$$

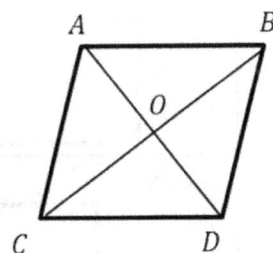

Rhombus for example.

Parallelogram Summary:

The following table summarizes what you need to prove a four-sided polygon is a particular type of quadrilateral.

Object	Criteria
Parallelogram 	All Opposite Sides Parallel OR All Opposite Sides Equal OR One Pair of Opposite Sides Equal and Parallel OR Opposite Angles Equal OR Adjacent Angles Supplementary OR Diagonals Bisect Each Other
Rhombus 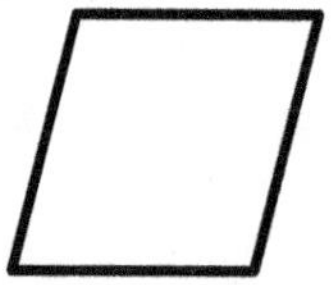	Prove that it's a Parallelogram AND Adjacent Sides Equal OR Diagonals Perpendicular OR Diagonals Bisect Vertex Angles
Rectangle 	Prove that it's a Parallelogram AND At Least One Angle is 90° OR Congruent Diagonals
Square	Prove that it's a Rectangle AND Adjacent Sides Equal OR Prove that it's a Rhombus AND At Least One Angle is 90°
Trapezoid	Opposite Sides Parallel

Chapter 9: Trapezoids

Isosceles Trapezoid Properties:

The isosceles trapezoid has these interesting properties:

> **For an isosceles trapezoid:**
> **Upper base angles are congruent,**
> **Lower base angles are congruent,**
> **Legs are congruent, and Diagonals are congruent.**

Given: The figure to the right is an isosceles trapezoid.

Prove: $\angle A \cong \angle B$; $\angle C \cong \angle D$; $\overline{AD} \cong \overline{BC}$

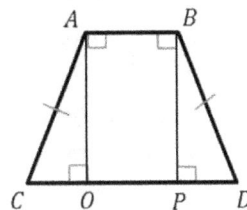

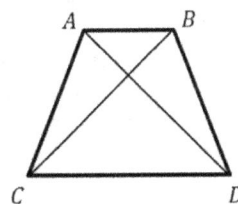

Trapezoid for proof.

Statement	Justification
$\overline{AC} \cong \overline{BD}$	Definition of Isosc. Trap.
$\angle C \cong \angle D$	Isosceles Triangle as shown
$\angle AOC \cong \angle BPD$	Perpendicular Lines
$\triangle AOC \cong \triangle BPD$	AAS Congruency
$\angle CAO \cong \angle PBD$	Triangle Congruency
$m\angle A = 90 + m\angle CAO$	Angle Addition
$m\angle B = 90 + m\angle PBD$	Angle Addition
$m\angle B = 90 + m\angle CAO = m\angle A$	Substitution
$\overline{AC} \cong \overline{BD}; \angle A \cong \angle B$	Given and Proven
$\overline{AB} \cong \overline{AB}$	Reflexive
$\triangle BAD \cong \triangle ABC$	SAS Congruency
$\overline{AD} \cong \overline{BC}$	Triangle Congruency

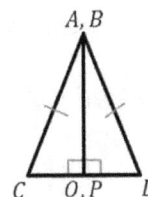

End triangles
pushed together.

General Trapezoid Properties:

A **median** of a trapezoid, a line connecting the midpoints of the legs, obeys the following property regardless of whether the trapezoid legs are equal or not:

> **The median of a trapezoid is parallel to its bases and has a length that is the average of the base lengths.**

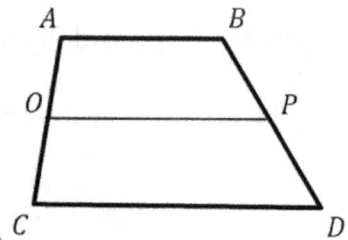

Trapezoid with median.

Given: $\overline{AO} \cong \overline{OC}; \overline{BP} \cong \overline{PD}; \overline{AB} \parallel \overline{CD}$
Prove: $\overline{OP} \parallel \overline{CD}; \ OP = (AB + CD)/2$

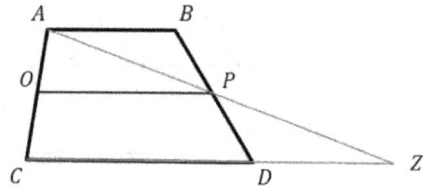

Figure for proof.

Statement	Justification
$\overline{BP} \cong \overline{PD}$	Given
$\angle BAP \cong \angle PZD$	Alt. Interior Angles with Parallel Lines & Transversal
$\angle ABP \cong \angle PDZ$	
$\triangle ABP \cong \triangle PDZ$	AAS Congruency
$\overline{AP} \cong \overline{PZ}; \overline{AB} \cong \overline{DZ}$	Triangle Congruency
$\overline{OP} \parallel \overline{CZ}$	$\overline{OP}$ is midsegment of $\triangle ACZ$
$\overline{OP} \parallel \overline{CD}$	$\overline{CD}$ is a segment of $\overline{CZ}$
$OP = \frac{1}{2}CZ = \frac{1}{2}(CD + DZ)$	$\overline{OP}$ is midsegment of $\triangle ACZ$
$OP = \frac{1}{2}(AB + CD)$	Substitution

> **Example**

For a 45-45-90 triangle with leg length equal to 10", find the length of a line drawn parallel to and one fourth up from one of the legs.

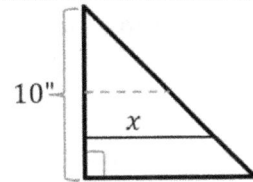

Triangle for example.

Since this is a 45-45-90 triangle, we know that the sides are equal, so the bottom of the triangle shown has length 10".

We draw a dotted line half way up as a midsegment to the triangle. By our first triangle midsegment theorem, we know the dotted line is parallel to the bottom leg and half its length, i.e. 5".

Finally, we discard the triangle from the dotted line up, and consider the remaining trapezoid. The line labeled x is now a median to the trapezoid which is the average length of the two bases:

$$x = \frac{10+5}{2} = 7.5".$$

> **Example**

Show that if $\overline{AE} \cong \overline{FB}$ and $ABCD$ is a rectangle, then $\overline{CF} \cong \overline{ED}$.

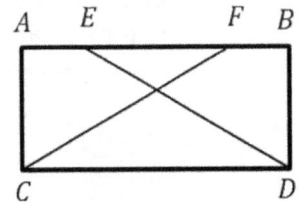

Figure for example.

We draw two dotted lines from E to C and F to D as shown. Because $ABCD$ is a rectangle, we know:

$$\overline{AB} \parallel \overline{CD}; \angle A \cong \angle B; \overline{AC} \cong \overline{BD}.$$

We are also given that $\overline{AE} \cong \overline{FB}$, so by SAS:

$$\triangle AEC \cong \triangle BFD \rightarrow \overline{EC} \cong \overline{FD}.$$

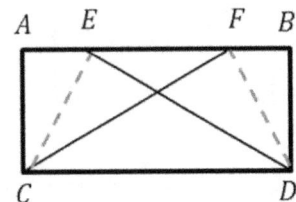

Figure redrawn.

By definition then, trapezoid $EFCD$ is isosceles, and $\overline{CF}$ and $\overline{ED}$ are its diagonals. Since diagonals of an isosceles trapezoid are congruent:

$$\overline{CF} \cong \overline{ED}.$$

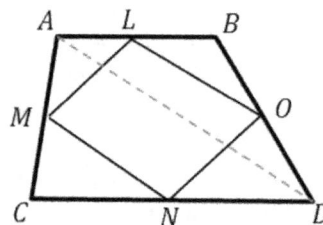

Trapezoid for example.

Example → Show that the midsegments of a trapezoid form a parallelogram.

We draw a dotted line to split the trapezoid into two triangles as shown. Since $\overline{MN}$ is a midsegment for the trapezoid, it must also be for the lower triangle. From this we know:

$$\overline{MN} \parallel \overline{AD}; \quad MN = \frac{AD}{2}.$$

Likewise, $\overline{LO}$ is a midsegment for the upper triangle:

$$\overline{LO} \parallel \overline{AD}; \quad LO = \frac{AD}{2}.$$

The two line segments are parallel to the same line and equal to the same measure, so:

$$\overline{MN} \parallel \overline{LO}; \quad MN = LO.$$

We have two opposing sides of the quadrilateral that are parallel and congruent. This is all we need to prove that $LMNO$ is a parallelogram.

Chapter 10: Circles

Definitions:

Circle: A collection of all planar points equidistant from a
center point. Notation: $\odot O$

Center: The point inside a circle from which all points on
the circle perimeter, or edge, are equidistant.

Diameter: A line segment that connects two points on the
perimeter and crosses the center point.

Radius: Any line segment from circle center to the
perimeter – it measures half the diameter.

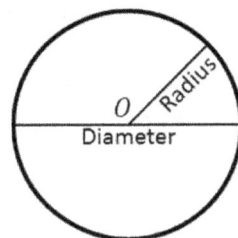

A circle.

> **Circumference:** The perimeter, p, of a circle:
> $$p = 2\pi r; \, r \text{ is the radius.}$$
> **Area:** The measure of the interior of a circle:
> $$area = \pi r^2 \, ; \, r \text{ is the radius.}$$

Arc: The portion of the circle between two points on the
perimeter. Notation: $\overset{\frown}{AB}$ measured in rad or degrees.

Minor/Major Arc: The minor arc is the small arc between
two points; the major arc is the larger arc with measure
360° minus the measure of the minor arc.

Sector: The portion defined by two radii and an arc.

Chord: A line segment connecting two points on the
perimeter. A diameter is a special chord.

Segment: The portion defined by a chord and its arc.

Secant: A line that crosses the circle in two points.

Tangent: A coplanar line or shape that crosses the circle at
exactly one point.

Concentric Circles: Circles that share the same center.

Inscribed Objects within a Circle: Objects whose vertices
are points on the perimeter of a containing circle.

Circumscribed Objects of a Circle: Objects whose sides are
tangent to the circle they contain.

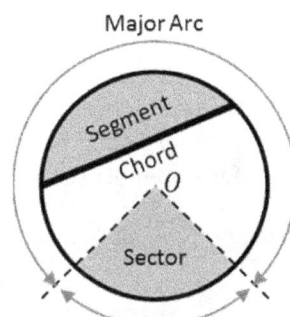

Arcs, chords, segments,
and sectors.

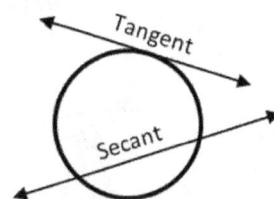

Tangents and secants.

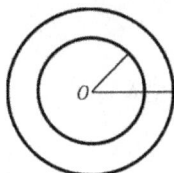

Concentric circles.

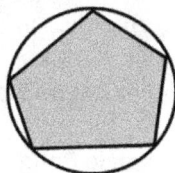

Inscribed polygon.

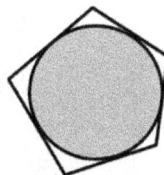

Circumscribed polygon.

Chords & Arcs:

There is a strong relationship between chords and the arcs they define. A theorem states:

> **The following statements are equivalent**
> **for the same or congruent circles:**
> - **Two chords are equidistant from the center**
> - **Two chords are congruent**
> - **Two arcs defined by such chords are congruent**

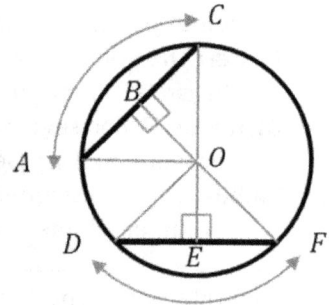

Two chords in a circle.
The grey triangles help prove:
$OB = OE \leftrightarrow AC = DF \leftrightarrow$
$\angle AOC \cong \angle DOF$.

By saying that the statements are equivalent, we have a three pronged if and only if statement. We leave the proof for Appendix D. By direct extension of the above theorem:

> **If two arcs or chords are congruent**
> **on the same or congruent circles**
> **Then the sectors and segments defined by them**
> **are each congruent.**

Another theorem on chords and arcs states:

> **If a diameter (or radius) is perpendicular to a chord,**
> **It bisects the chord and its arc.**

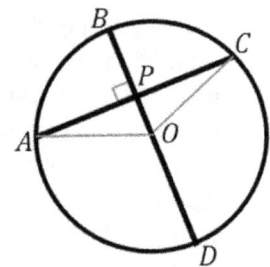

Diameter perpendicular
to a chord.

Given: $\overline{AC} \perp \overline{BD}$
Prove: $\overset{\frown}{AB} \cong \overset{\frown}{BC}$ ($\angle BOA \cong \angle BOC$); $\overline{AP} \cong \overline{PC}$

Statement	Justification
$\overline{OP} \cong \overline{OP}$; $\overline{OA} \cong \overline{OC}$	Reflexive; Radii
$\angle APO \cong \angle CPO$	Perpendicular Lines
$\triangle APO \cong \triangle CPO$	Right Angle Exception to SSA
$\angle BOA \cong \angle BOC$; $\overline{AP} \cong \overline{PC}$	Triangle Congruency

With this we can state the final theorem of this section:

> **If two chords are parallel
> the arcs between them are congruent.**

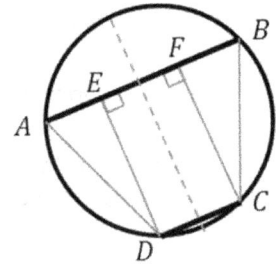

Parallel chords.

If we draw rectangle $CDEF$, we know that $ED = FC$ and $EF = CD$. Drawing a diameter that bisects $\overline{CD}$ and $\overline{AB}$, we can use the prior theorem an d a little algebra to show that $AE = FB$.

With two right triangles, by SAS $\triangle AED \cong \triangle BFC$, which implies $AD = BC$. Finally, because the chords are equal, the arcs must also be equal.

Example

A chord is 4" from the center of circle of radius 5". Find the length of the chord.

A perpendicular drawn to the chord from the center of the circle will bisect the chord. Drawing a radius to the chord endpoint completes a right triangle. Using Pythagorean's theorem:

$$c^2 = 5^2 = a^2 + b^2 = 4^2 + \left(\frac{x}{2}\right)^2$$

$$\frac{x}{2} = \sqrt{25 - 16} = \sqrt{9} = 3$$

$$x = 6"$$

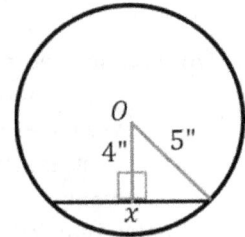

Chord for example.

Example

Show that $\overline{AC} \cong \overline{BD}$ if $\overline{AB} \parallel \overline{CD}$.

Since chords $\overline{AB}$ and $\overline{CD}$ are parallel, then their arcs are equal: $\overset{\frown}{AD} \cong \overset{\frown}{BC}$. Since the arcs are equal, so are their respective chords: $AD = BC$.

Now we have two parallel "bases" and two equal legs which means $ABCD$ is an isosceles trapezoid. The diagonals of an isosceles trapezoid are equal, therefore $\overline{AC} \cong \overline{BD}$.

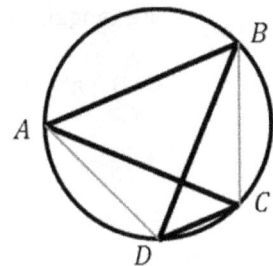

Chords for example.

Tangents & Secants:

There are two theorems associated with tangents:

> **A line through a point of a circle is a tangent iff it is perpendicular to the radius drawn to that point.**

As an informal or intuitive proof, consider the chord $\overline{RS}$. We draw a diameter or radius perpendicular to the chord to bisect it. As $\overline{RS}$ moves further away from circle center, the length of it on either side of the radius evenly decreases. $\overline{RS}$ eventually becomes so small as to approach becoming a single point where the radius meets the circle edge. The line through that single point, and no other point on the circle, is by definition a tangent.

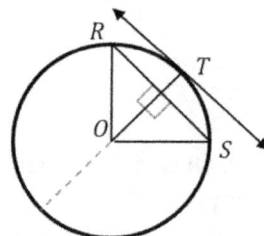
Radius perpendicular to chords and the tangent.

Our second theorem states:

> **If two tangents intersect, the segments from the intersection to the circle are congruent.**

We know that $OT = OU$ (radii), $OP = OP$ (reflexive), and $\angle PTO \cong \angle PUO$ (above theorem), so by the exception to SSA for right triangles we have $\triangle PTO \cong \triangle PUO$, and therefore $PT = PU$.

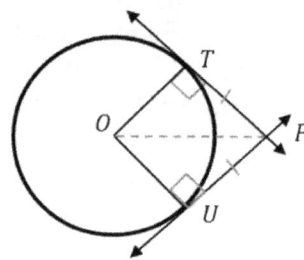
Two tangents to a circle.

> **Example**
>
> Show $\overline{RS} \cong \overline{TU}$ if the lines shown through point Q are tangent to both circles.
>
> By the above theorem,
>
> $$QR = QT \text{ and } QS = QU.$$
>
> By vertical angles, $\angle RQS \cong \angle TQU$.
>
> So, by SAS $\triangle RQS \cong \triangle TQU$, and $RS = TU$.

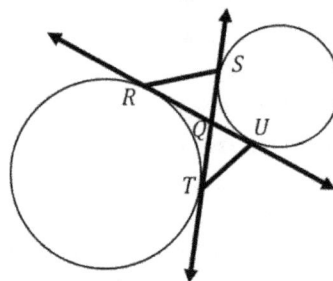
Figure for example.

Areas & Angles:

As called out in the definitions:

> **The area of a circle is πr^2.**
> **The circumference of a circle is $2\pi r$.**

where r is the radius, and π is 3.14 rad. A sector is a portion of a circle so its area and circumference are calculated as a percentage of a whole circle:

> **The area of a sector is $\left(\frac{n}{360°}\right)\pi r^2$.**
> **The circumference traced out by a sector is $\left(\frac{n}{360°}\right)2\pi r$.**

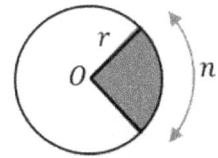

Circle & sector.

where n is the angular measure of the arc of the sector.

There are three types of angles that can be measured relative to a circle. For **inscribed angles** (vertex on the circle):

> **An inscribed angle measures**
> **half that of the intercepted arc.**

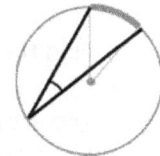

Inscribed angle.

For an **interior angle** (vertex inside of the circle):

> **An interior angle formed by**
> **crossing lines interior to the circle measures**
> **half the sum of its intercepted arcs.**

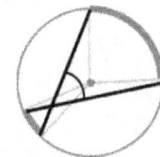

Interior angle.

For an **exterior angle** (vertex outside of the circle):

> **An exterior angle measures**
> **half the difference of its intercepted arcs.**

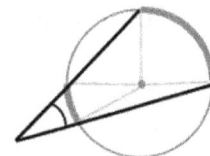

Exterior angle.

Let's look at some examples.

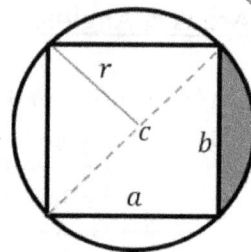

Square inscribed
in circle.

Example A square is inscribed in a circle of radius 5". Find the area of the shaded region.

The diagonal of the square matches the diameter of the circle, or $2r = 10$". The bottom triangle is an isosceles right triangle, so $a = b$ in Pythagorean's equation:

$$c^2 = 10^2 = 100 = a^2 + b^2 = 2a^2$$

$$a = \sqrt{100/2} = \frac{10}{\sqrt{2}}\text{ "}$$

The area of the circle is $2\pi r^2 = 2(25)\pi = 157$ in²
The area of the square is $a^2 = \frac{100}{2} = 50$ in²

The area of the shaded region is ¼ the difference of the areas of the circle and square:
$$A = \frac{157-50}{4} = 26.75 \text{ in}^2$$

Example Find the measure of the exterior angle.

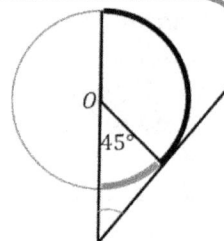

Exterior angle for
example.

The arc in grey is given as 45°. The black arc is half the circle minus the grey arc: $180° - 45° = 135°$.

The exterior angle is then:

$$\frac{1}{2}(135° - 45°) = 45°$$

Example Find an expression for x and y in terms of a and b.

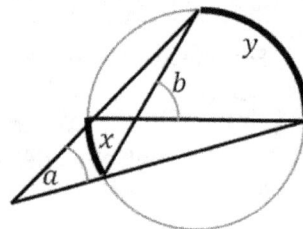

Figure for example..

As a is an exterior angle: $a = (y - x)/2$
As b is an interior angle: $b = (y + x)/2$

Adding equations: $a + b = \frac{y-x+y+x}{2} = y$

Subtracting equations: $b - a = \frac{y+x-y+x}{2} = x$

Chapter 11: Transformations

Symmetry & Transformations:

Symmetry is achieved in an object when one part of an object duplicates another. There are three types of symmetry:

Line Symmetry: Reflection about a line of symmetry.

Rotational Symmetry: Rotation about a point. A regular polygon has $360°/n$ symmetry. For example, an octagon can be rotated $360°/8 = 45°$ to duplicate itself.

Point Symmetry: $180°$ rotation about a point.

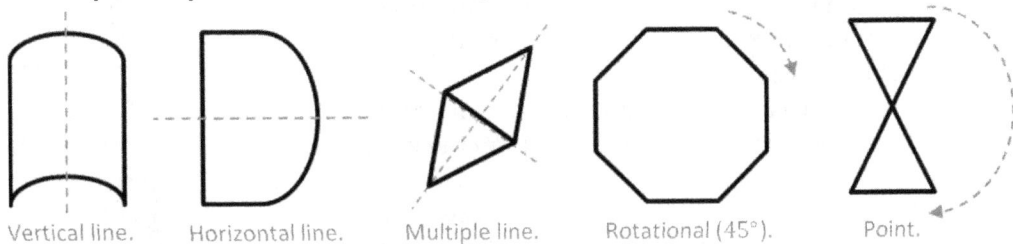

Vertical line.   Horizontal line.   Multiple line.   Rotational (45°).   Point.

A **transformation** takes the points of an original object, a **preimage**, and forms them into another object, an **image**. We say there is a **1:1 mapping** of the points of the preimage to the image meaning that every point is accounted for and treated in the same manner. We write $A \rightarrow A'$, meaning preimage A is transformed to image A'.

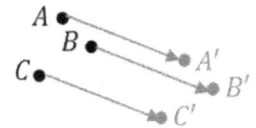

1:1 mapping for all points.

The **orientation** of an object is the clockwise (CW) or counter-clockwise (CCW) tracing of an image around its perimeter. If you choose to draw a preimage in one orientation, and its image can be drawn in the same order and orientation, then there is a **direct transformation**. If the orientation is reversed, then there is an **opposite transformation**.

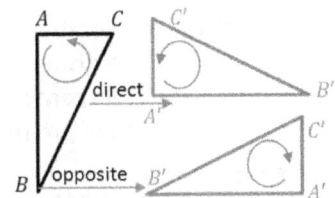

Direct versus opposite transformation. The curved arrows show drawing the image from A to B to C.

Transformations can also be categorized by their ability to preserve congruence. When a preimage and image are congruent, there is a **congruence transformation**, or **isometry**. If the image is a scaled replica of the preimage, it is a **similar transformation**.

There are three basic isometries (congruent transforms):

Reflection: An opposite transformation that flips an object around a line, L, or point, P. Notation: $r_L(A) = A'$

Reflection: Each point is equidistant from the line.

> The perpendicular distance between each preimage point to the line or point of isometry is equal to the perpendicular distance between each image point and the line or point of isometry.

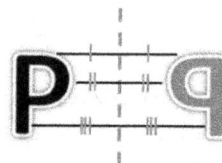

Translation: A direct transformation that slides an object h units to the right and k units upward (a negative h or k moves left or down, respectively). Notation: $T_{h,k}(A) = A'$

Translation: Each point is slid over in exactly the same way.

> Each point of the preimage is translated in precisely the same way to form the image.

Rotation: A direct transformation that spins an object counterclockwise $a°$ about a point called the **center of rotation** (CoR). Notation: $R_{a°}(A) = A'$

Rotation: Each segment pair forms the same angle.

> The distance between each point of the preimage to the CoR is the same as the distance of each point of the image to the CoR.

There is only one similar transformation:

Dilation: A direct transformation that scales an object by a factor of c, called the **scale factor** or **constant of dilation**. Notation: $D_c(A) = A'$

Dilation: A scaled replica.

> The distance of each point of the image relative to its center is c times the distance of each point of the preimage to its center.

Example

Name the transformation:

Translation   Rotation   Dilation   Reflection

Composite Transformations:

You can do multiple transformations one after another:

Composite Transformation: Performing a series of transformations in order, from right to left. For example: $R_{30°} \circ T_{2,3}(A) = A'$ or $R_{30°}\big(T_{2,3}(A)\big) = A'$ means translate, then rotate.

Translation and rotation.

Glide Transformation: A special opposite transformation that reflects about a line and translates parallel to that line. Order is not important for this case: $r_L \circ T_{h,k}(A) = T_{h,k} \circ r_L(A)$

Glide transformation.

There are some interesting properties about multiple reflections. If you perform two translations you end up with one big translation, and two rotations just gives you one big rotation. But two reflections give you:

- About two parallel lines: A translation
- About two intersecting lines: A rotation

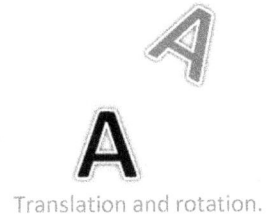

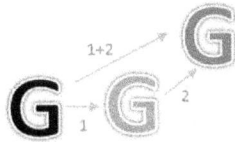

Two translations are one big translation.

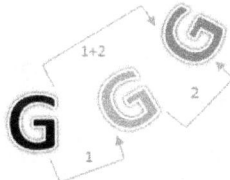

Two rotations are one big rotation.

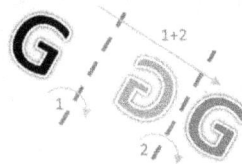

Two reflections can result in one big translation or one big rotation.

This means that any single congruent transformation can be expressed as one or two reflections. And, any two congruent objects can be thought of as no more than three reflections of one object relative to the other.

Example

Show that a glide is three reflections.

From the picture, we reflect once over the positive sloped line, then twice over the parallel lines that are perpendicular to the first line of isometry.

A glide is 3 reflections.

Transformations with Coordinates:

Let's look at each basic transformation in the context of a coordinate system. The x-coordinate is called the **abscissa**, and the y-coordinate is called the **ordinate**. We'll refer to preimage points (A_x, A_y) and image points (A'_x, A'_y).

Reflection:

About a horizontal line, $r_{x=a}(A_x, A_y)$: If given the line $x = a$, a reflection about this line is:
$$(A'_x, A'_y) = ([2a - A_x], A_y).$$

About a vertical line, $r_{y=b}(A_x, A_y)$: If given the line $y = b$, a reflection about this line is:
$$(A'_x, A'_y) = (A_x, [2b - A_y]).$$

About a point: If reflecting about a point (a, b), $r_{(a,b)}(A_x, A_y)$:
$$(A'_x, A'_y) = ([2a - A_x], [2b - A_y]).$$

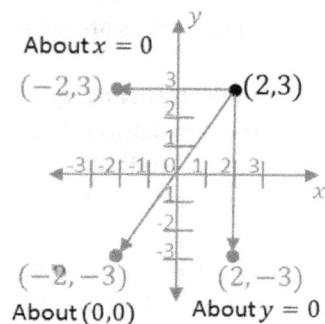

Reflections.

Translation:

To translate a point h units horizontally and k units vertically, $T_{h,k}(A_x, A_y)$:
$$(A'_x, A'_y) = ([A_x + h], [A_y + k]).$$

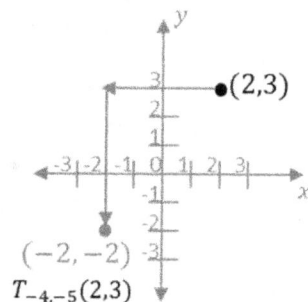

Translation.

Rotation:

To rotate CCW *about the origin* use these guidelines, $R_{angle°}(A_x, A_y)$:

$0°, 360°$:	$(A'_x, A'_y) = (A_x, A_y)$;
$90°$:	$(A'_x, A'_y) = (-A_y, A_x)$;
$180°$:	$(A'_x, A'_y) = (-A_x, -A_y)$;
$270°, -90°$:	$(A'_x, A'_y) = (A_y, -A_x)$.

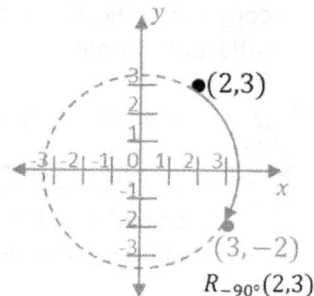

Rotation.

Dilation:

To dilate by a factor of c *about the origin*, $D_c(A_x, A_y)$:
$$(A'_x, A'_y) = ([cA_x], [cA_y]).$$

Perform the following transformations on the line given to the right: $r_{x=-1}$; $T_{-2,1}$; $R_{-90°}$; D_2

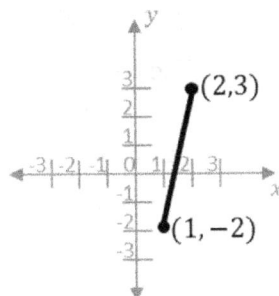

Line for example.

For each transformation, we will transform the two endpoints then connect them with a line.

$r_{x=-1}$: $(A'_x, A'_y) = ([2a - A_x], A_y)$.

The coordinate $(2,3)$ transforms to
$$(2 \times (-1) - 2, 3) = (-4, 3).$$

The coordinate $(1, -2)$ transforms to
$$(2 \times (-1) - 1, -2) = (-3, -2).$$

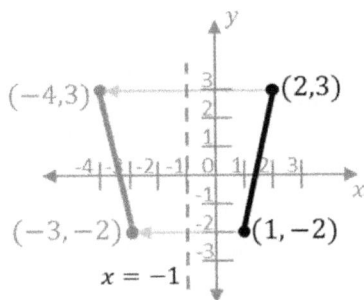

$T_{-2,1}$: $(A'_x, A'_y) = ([A_x + h], [A_y + k])$.

The coordinate $(2,3)$ transforms to
$$(2 - 2, 3 + 1) = (0, 4).$$

The coordinate $(1, -2)$ transforms to
$$(1 - 2, -2 + 1) = (-1, -1).$$

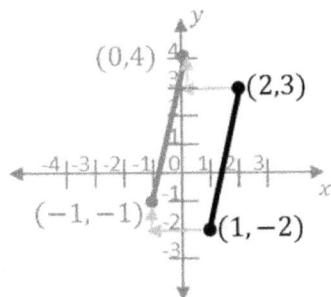

Ex.Cont.

$R_{-90°}$: $(A'_x, A'_y) = (A_y, -A_x)$.

The coordinate $(2,3)$ transforms to
  $(3, -2)$.

The coordinate $(1, -2)$ transforms to
  $(-2, -1)$.

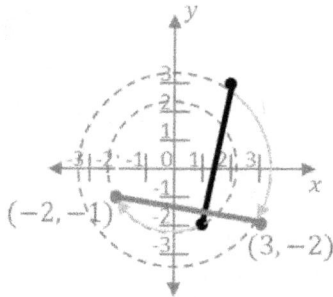

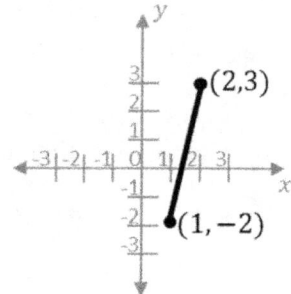

Line for example.

D_2: $(A'_x, A'_y) = ([cA_x], [cA_y])$.

The coordinate $(2,3)$ transforms to
  $(2 \times 2, 2 \times 3) = (4,6)$.

The coordinate $(1, -2)$ transforms to
  $(2 \times 1, 2 \times -2) = (2, -4)$.

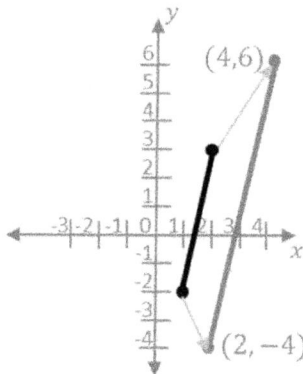

Chapter 12: Locus & Concurrency

Locus:

A **simple locus** is a set of points that satisfies a particular condition. A **compound locus** is a set of points that satisfies multiple conditions. The five loci worth noting are:

Equidistant From...	Locus	Locus Shown in Black
Single point	Circle	
Two points	Line (perpendicular bisector of line between the points)	
Single line	Parallel lines	
Parallel lines	Line (halfway between the parallel lines)	
Intersecting lines	Two lines (vertical angle bisectors)	

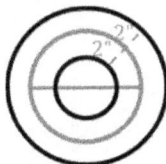

> **Example**
>
> Find the loci that are 2" from a circle of radius 4" and the horizontal diameter of that circle.
>
> Draw each locus, then find the intersections.
>
> Original Condition:
> 4" radius ⊙ & its horizontal diameter.
>
> Locus 1:
> All points 2" from circle.
>
> Locus 2:
> All points 2" from diameter.
>
> Final answer:
> Locus 1 ∩ Locus 2.

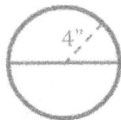

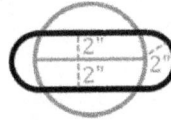

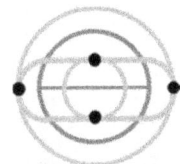

Concurrency:

When more than two lines intersect at the same point, they are called **concurrent lines**. The intersection point is called the **point of concurrency**. For a triangle, there are four sets of concurrent lines:

The medians: The medians intersect at a point called the **centroid** or **center of gravity**. The distance from a vertex to the centroid is 2/3 the length of its median.

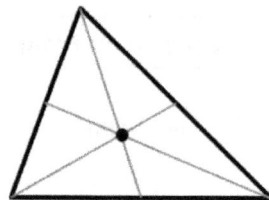

Centroid.

The altitudes: The altitude intersection point is called the **orthocenter**. For acute triangles, the orthocenter is interior to the triangle, for right triangles it is at the right angle vertex, and for obtuse triangles it is exterior to the triangle.

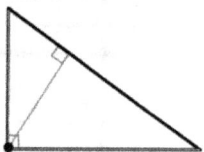

Right triangle.

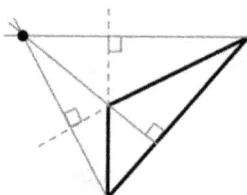

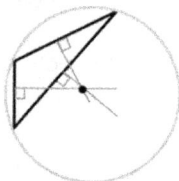

Obtuse triangle.

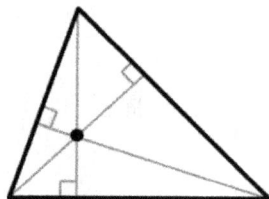

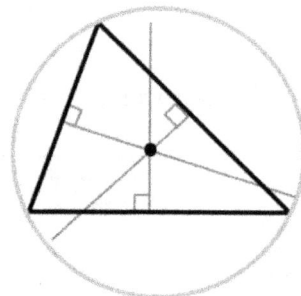

Orthocenter.

The perpendicular side bisectors: The intersection of perpendicular side bisectors is called the **circumcenter**. This is because it is equidistant to all vertices meaning it is also the center of a circumscribing circle. For acute triangles, the circumcenter is interior to the triangle, for right triangles it is on the hypotenuse, and for obtuse triangles it is exterior to the triangle.

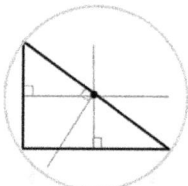

Right triangle.

Obtuse triangle.

Circumcenter.

The vertex bisectors: The intersection of vertex bisectors is called the **incenter** because it is equidistant from all sides and the center of an inscribed circle.

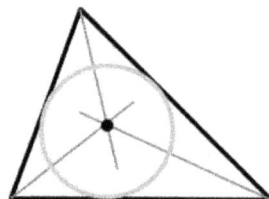

Incenter.

Example

Show that the diagonals and perpendicular side bisectors of a rectangle are lines of concurrency.

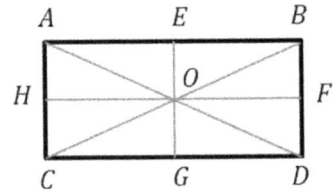

Rectangle for example.

The problem is asking whether the diagonals and perpendicular bisectors cross at the same point. We can see $\overline{AD}$ and $\overline{BC}$ are diagonals, so we must determine if $\overline{OG}, \overline{OH}, \overline{OE},$ and $\overline{OF}$ are perpendicular side bisectors

We know that diagonals of a rectangle are congruent and bisect each other from Chapter 8. Looking at ΔCOD, we have an isosceles triangle with an altitude $\overline{OG}$. Since $\overline{OG} \perp \overline{CD}$ and $\overline{OC} \cong \overline{OD}$, then $\overline{OG}$ is a perpendicular bisector of $\overline{CD}$. Making the same arguments all around, $\overline{OH}, \overline{OE},$ and $\overline{OF}$ are also perpendicular side bisectors, which do intersect the diagonals at a point of concurrency.

Example

Show that perpendicular bisectors of two nonparallel chords of a circle intersect at the circle center.

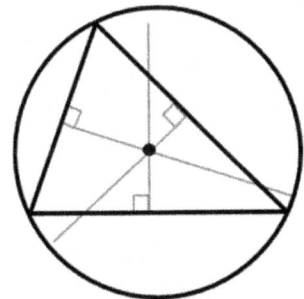

Circumcenter.

Looking at the figure for a circumcenter, we see that any two legs of the triangle satisfy the condition of being nonparallel chords of a circle. But, let's make the figure a bit more general by considering just two chords rather than a whole triangle.

We draw a concentric circle (from the same center) so that one edge of the triangle is exterior. The endpoints of the new chords are all a radius distance from center. Since any point on a perpendicular bisector is equidistant to the endpoints of the line segment (Chapter 5), the lines shown from center to each chord midpoint are perpendicular bisectors of those chords.

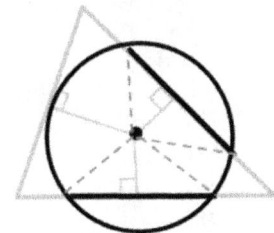

Relationships with concentric circle. Dashed lines are radii.

Chapter 13: Solids

Cylinders & Prisms:

A **solid** is a is a three dimensional (3-D) object. A **cylinder** is a solid formed by connecting two circles in parallel planes. The circles are called **bases**, and the side joining the two circles is the **lateral face**. If the line joining the centers of the two circles is perpendicular to the bases, the cylinder is **right**; if it is slanted, the cylinder is **oblique**. The **height** of a cylinder is the perpendicular distance between bases.

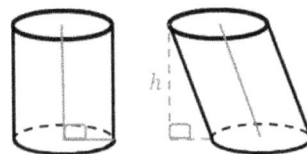

Right and oblique cylinders.

A **polyhedron** is a 3-D object with polygons for faces. A **prism** is a polyhedron formed by connecting two identical polygons in parallel planes. The polygons are the **bases** whose perimeters are **edges** and corners are **vertexes**. The sides connecting the two polygons are the **lateral faces**, and the intersection of lateral faces is a **lateral edge**. A special case is the **cube** which is two square bases connected by four square faces. Just as with cylinders, prisms can be **right** or **oblique**, and **height** is measured as the perpendicular distance between the two bases.

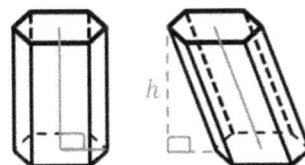

Right and oblique prisms.

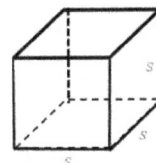

Cube.

The **volume** of a cylinder or prism is the space the 3-D object occupies. The **lateral area** is the area of all of the lateral sides. And, the **surface area** is the sum of the lateral area and the areas of the bases .

With V = volume, B = base area, F = face area, S = surface area, and p = the base perimeter or circumference, we have the following formulas:

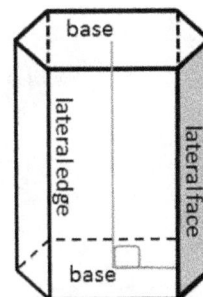

Definitions.

> **For any cylinder or prism:**
>
> $$V = Bh$$
>
> **For any right cylinder or right prism:**
>
> $$F = ph; \quad S = 2B + F$$

Find the volume of a cylinder with radius 4" and side length, $s = 10$", that slants 45°.

The area of the circles is:

$$B = \pi r^2 = 16\pi \text{ in}^2$$

For a 45° slant, the height, h, is given by:

$$c^2 = 10^2 = a^2 + b^2 = 2h^2 \rightarrow h = \frac{10}{\sqrt{2}} = 7.1 \text{ in.}$$

Plugging into our formulas:

$$V = Bh = (16\pi)(7.1) = 356.9 \text{ in}^3$$

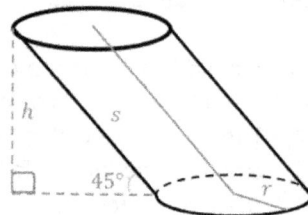

Cylinder for example.

Find the volume, lateral surface area, and surface area of right 8" high prism formed with 2" × 2" squares.

The perimeter and area of the bases are:

$$p = 4s = 8 \text{ in}$$
$$B = s^2 = 4 \text{ in}^2$$

Plugging into our formulas:

$$V = Bh = (4)(8) = 32 \text{ in}^3$$
$$F = ph = (8)(8) = 64 \text{ in}^2$$
$$S = 2B + F = 8 + 64 = 72 \text{ in}^2$$

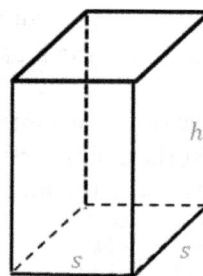

Prism for example.

Cones & Pyramids:

A **cone** is a 3-D object formed by connecting the perimeter points of a circle in one plane to a single point in another. The circle is the **base**, the point is the **vertex**, and the **height** is the perpendicular distance from the vertex to the circle plane. Cones can be **right** or **oblique** if the line drawn between the vertex and circle center is perpendicular or nonperpendicular to the circle plane, respectively. For a right cone, the **slant height** is the length of a line from the vertex to the circle edge.

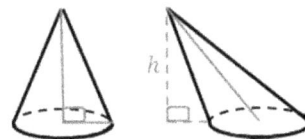

Right and oblique cones.

A **pyramid** is a polyhedron formed by connecting the perimeter of a polygon in one plane to a point in another. The polygon is the **base**, the point is the **vertex**, the sides are the **lateral faces**, and the side edges are the **lateral edges**. Pyramids can be **right** or **oblique**, and **height** is measured as the perpendicular distance between the vertex and the base. For a **regular pyramid** (a right pyramid with a regular polygon for a base), the **slant height** is the length from the vertex to the middle of a side of the base.

Right and oblique pyramids.

The **volume** of a cone or pyramid is the space the 3-D object occupies. The **lateral area** is the area of all of the lateral sides. And, the **surface area** is the sum of the lateral area and the area of the base . With V = volume, B = base area, F = face area, S = surface area, and p = the base perimeter or circumference, we have the following formulas:

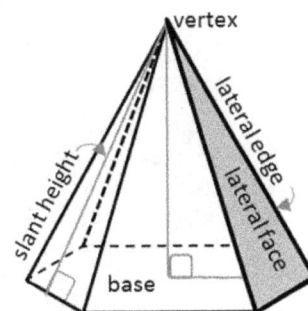

Definitions.

> **For a cone or pyramid:**
>
> $$V = \frac{1}{3}Bh$$
>
> **For any right cone or right pyramid:**
>
> $$F = \frac{1}{2}ph; \quad S = B + F$$

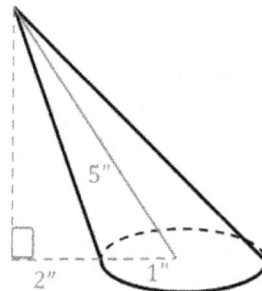

Cone for example.

Example

Find the volume of the cone shown.

The area of the circle is:

$$B = \pi r^2 = \pi \text{ in}^2$$

For the height of a right triangle with side 3" and hypotenuse 5":

$$c^2 = 5^2 = a^2 + b^2 = 3^2 + h^2 \rightarrow h = 4 \text{ in.}$$

Plugging into our formulas:

$$V = \frac{1}{3}Bh = \frac{1}{3}(\pi)(4) = 4.2 \text{ in}^3.$$

Example

Find V, F, and S of a regular pyramid with a base hexagon of side 2" and a height of 5".

The perimeter is simply $p = 6 \times 2 = 12$ in. To find the area, we need to find the apothem, AC:

$$\angle BAD = \frac{360°}{6} = 60° \rightarrow \angle BAC = \frac{60°}{2} = 30°.$$

For a 30-60-90 triangle, the relationship of sides is $1:\sqrt{3}:2$. The short side, $\overline{BC}$ is 1", therefore $AC = \sqrt{3}$". The area is then:

$$B = \frac{ap}{2} = AC \times \frac{p}{2} = \sqrt{3} \times 6 = 10.4 \text{ in}^2$$

Plugging into our formulas:

$$V = \frac{1}{3}Bh = \frac{1}{3}(10.4)(5) = 52 \text{ in}^3$$

$$F = \frac{1}{2}ph = \frac{1}{2}(12)(5) = 30 \text{ in}^2$$

$$S = B + F = 10.4 + 30 = 40.4 \text{ in}^2.$$

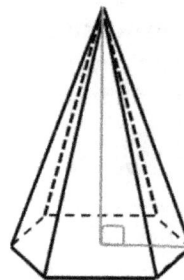

Pyramid for example.

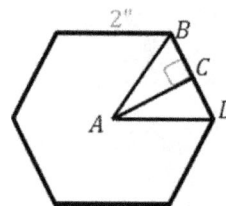

Base of pyramid.

Spheres:

A **sphere** is a 3-D object formed by all points equidistant from a common **center**. The measure from the center to the perimeter of the sphere is the **radius**. The sphere can be cut in half, forming **hemispheres**, by splitting the sphere along its largest cross-section. The largest cross-section is also referred to as the **great circle**.

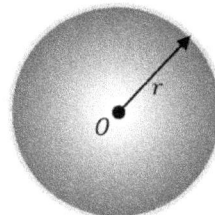

A sphere.

The **volume** of a sphere is the space the 3-D object occupies. The **surface area** is the area taken up by the sphere perimeter if it were opened up to lay flat. With $V =$ volume, $S =$ surface area, and $r =$ the radius, we have the following formulas:

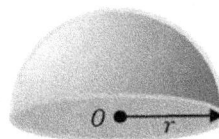

A hemisphere.

For a sphere:

$$V = \frac{4}{3}\pi r^3$$

$$S = 4\pi r^2$$

Example

Find V and S for a sphere of radius 5".

$$V = \frac{4}{3}\pi r^3 = \frac{4}{3}\pi 5^3 = 167\pi = 524 \text{ in}^3.$$

$$S = 4\pi r^2 = 4\pi 5^2 = 100\pi = 314 \text{ in}^2.$$

Appendices

A: Course Summary
B: Problem Sets
C: Solutions to Problem Sets
D: Additional Proofs
E: Introduction to Trigonometry

Appendix A: Course Summary

Overview of basics:

1
    Definitions for points, lines, planes, angles
    Tools – ruler, protractor, compass
    Theorems, corollaries, postulates

2
    Logic statements and logic operators
    Starting postulates

Important Angles:

3
    Vertical angles $\cong$
    Parallel lines cut by a transversal produce
        $\cong$alt. interior angles
        $\cong$alt. exterior angles
        $\cong$corresponding angles

Triangles:

4
    Sum of interior angles is 180°
    Exterior angle is sum of other 2 interior angles
    Inequalities & relationships of sides
    $\overline{BC} < \overline{AC} \leftrightarrow \angle A < \angle B$

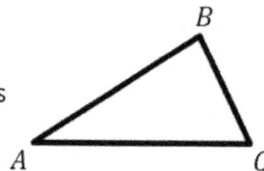

Congruent triangles

5
    Proven by SSS, SAS, ASA, AAS
    Points on perpendicular bisectors are equidistant from endpoints
    Points on angle bisectors are equidistant from angle sides
    Isosceles triangles : $\overline{BC} = \overline{AC} \leftrightarrow \angle A = \angle B$

Similar triangles

6
    Proven by AA, SSS, SAS
    Sides, altitudes, and perimeters are in same proportion
    Areas are in square proportion to sides
    Midsegments are parallel and half length of third side
    Midsegment triangles are ½ perimeter, ¼ area of original triangle

Right triangles

7
    Altitude makes 3 similar triangles
    Other angles are complementary
    Pythagorean's Theorem: $c^2 = a^2 + b^2$

Polygons:

Perimeter is sum of sides, Area of regular polygon $= ap/2$

Sum of interior angles is $180°/n$

Sum of exterior angles is $360°$

Parallelogram:

Opposite angles $\cong$; Opposite sides $\cong$

Adjacent angles supplementary

Diagonals bisect each other

Table for proving special quadrilaterals – p 44

Trapezoids:

Isosceles: Upper angles $\cong$, lower angles $\cong$, diagonals $\cong$

Any trapezoid: Median is parallel to and average length of bases

Circles:

Perimeter $= 2\pi r$; Area $= \pi r^2$

$\cong$ chords, $\cong$ arcs, $\cong$ distance from center

Radius perpendicular to chord bisects it & its arc

Radius drawn to tangent is perpendicular to it

Arcs between parallel chords are $\cong$

Two intersecting tangents have same segment length to circle

Angles in circles:

inscribed = ½ arc, interior = ½ sum of arcs, exterior = ½ difference of arcs

Symmetry/Transformations/Locus/Concurrency:

Symmetry: Line, rotational, & point

Transformations: Direct & opposite, congruent & similar

Translation, rotation, reflection, dilation, composites, and glides

Working in coordinates – p 58

Locus is a set of points satisfying certain conditions

Concurrency: 4 types for triangles – p 62

Solids:

	Cylinder, Prism	Cone, Pyramid	Sphere
Volume:	$V = Bh$	$V = \frac{1}{3}Bh$	$V = \frac{4}{3}\pi r^3$
Face Area (Regular)	$F = ph$	$F = \frac{1}{2}ph$	
Surface Area (Regular)	$S = 2B + F$	$S = B + F$	$S = 4\pi r^2$

Appendix B: Problem Sets

Chapter 1

Find the slope, midpoint, and distance for the following pairs of points:

1.1: $(1,2); (3,4)$

1.2: $(-1,-2); (3,4)$

1.3: $(-5,-3); (-3,0)$

1.4: $(4,8); (-5,2)$

Use a protractor to measure the angle (extend the lines if needed):

1.5:

1.6:

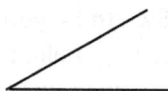

1.7:

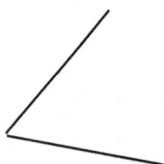

1.8:

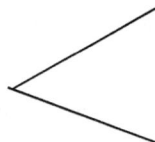

Find the perpendicular bisector using a compass:

1.9:

1.10:

1.11:

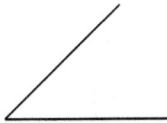

1.12:

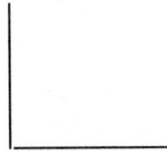

Find the angle bisector using a compass:

1.13:

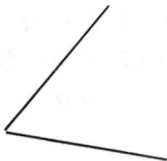

1.14:

1.15:

1.16:

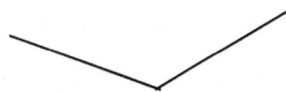

Chapter 2

Identify the kind of reasoning needed or demonstrated:

2.1: You have six pictures in front of you that look similar. You are told two of them are exact matches. How do you determine which two?

2.2: Your mother says your room is never clean, but today it is. How do you refute her statement?

2.3: You are asked to walk into a perfectly square room and measure all four walls. Why do you only measure one?

2.4: A neighbor tells you that a steady diet of junk food is bad for your weight. You eat 50 candy bars in a week and end up gaining 15 lbs. How did you prove the neighbor right?

Evaluate:

2.5: $p = q = True$
$(p \wedge \sim q) \vee (p \wedge q)$

2.6: $p = q = False$
$(p \wedge \sim q) \vee (p \wedge q)$

2.7: $p = q = True$
$(p \vee \sim q) \wedge (p \vee q)$

2.8: $p = q = False$
$(p \vee \sim q) \wedge (p \vee q)$

2.9: $p = \{3 > 2\}$;
$q = \{-3 > -2\}$
$(\sim p \wedge \sim q) \wedge (p \wedge q)$

2.10: $p = \{2 + 2 = 5\}$;
$q = \{2 \times 4 = 8\}$
$(\sim p \vee \sim q) \vee (p \vee q)$

2.11: $p = \{AB > AD\}$;
$q = \{AD = AC + BD\}$
$(\sim p \wedge q) \wedge \sim (p \wedge q)$

2.12: $p = \{AD > AB\}$;
$q = \{AD > AB + CD\}$
$(\sim p \wedge q) \wedge \sim (p \wedge q)$

2.13: $p = \{AD = AC + CD\}$;
$q = \{AD = AB + BD\}$
$\left(q \wedge (p \wedge q)\right) \vee \sim p$

2.14: $p = \{m\angle MLN = m\angle MLO\}$;
$q = \{m\angle MLN > m\angle MLO\}$
$\left(q \wedge (p \wedge q)\right) \vee \sim p$

2.15: $p = \{m\angle MLO > m\angle NLO\}$;
$q = \{m\angle NLO > m\angle MLO\}$
$\left(q \wedge (p \wedge q)\right) \vee \sim p$

2.16: $p = \{m\angle MLN < m\angle MLO\}$;
$q = \{m\angle NLO < m\angle MLO\}$
$\left(\sim q \wedge (p \wedge q)\right) \vee p$

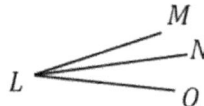

Chapter 3

Solve for x:

3.1:

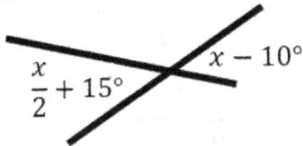

$$\frac{x}{2} + 15°$$

$$x - 10°$$

3.2:

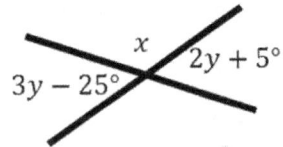

$$3y - 25°$$

$$x$$

$$2y + 5°$$

3.3:

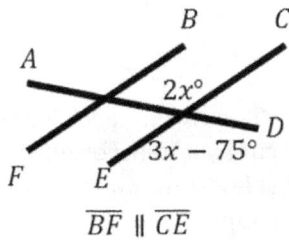

$$2x°$$

$$3x - 75°$$

$$\overline{BF} \parallel \overline{CE}$$

3.4:

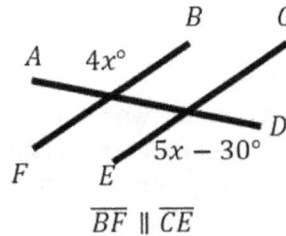

$$4x°$$

$$5x - 30°$$

$$\overline{BF} \parallel \overline{CE}$$

3.5:

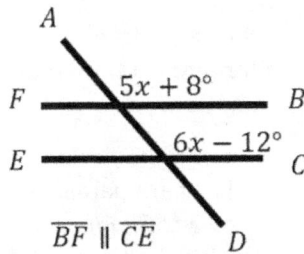

$$5x + 8°$$

$$6x - 12°$$

$$\overline{BF} \parallel \overline{CE}$$

3.6:

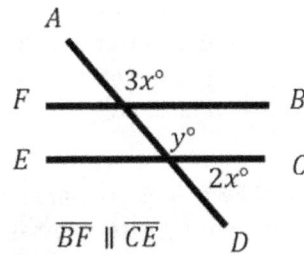

$$3x°$$

$$y°$$

$$2x°$$

$$\overline{BF} \parallel \overline{CE}$$

Determine if $a = b = c$:

3.7:

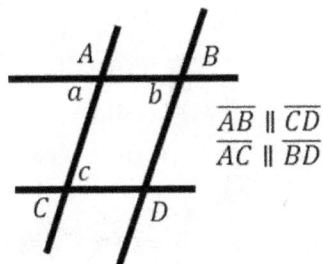

$$a \quad b$$

$$c$$

$$\overline{AB} \parallel \overline{CD}$$
$$\overline{AC} \parallel \overline{BD}$$

3.8:

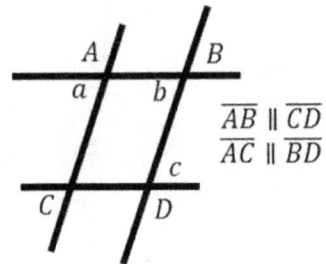

$$a \quad b$$

$$c$$

$$\overline{AB} \parallel \overline{CD}$$
$$\overline{AC} \parallel \overline{BD}$$

Determine if $\overline{AC} \parallel \overline{BD}$:

3.9:

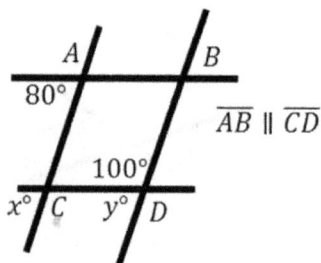

$\overline{AB} \parallel \overline{CD}$

3.10:

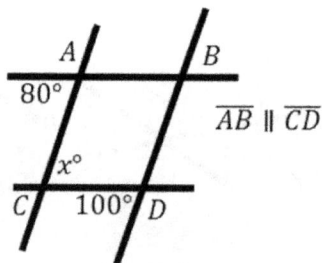

$\overline{AB} \parallel \overline{CD}$

Prove the following:

3.11: When two parallel lines are cut by a transversal, same side interior angles are supplementary.

3.12: When two parallel lines are cut by a transversal, same side exterior angles are supplementary.

3.13: Two lines are parallel if when cut by a transversal, a pair of alternate interior angles are congruent.

3.14: Two lines are parallel if when cut by a transversal, a pair of alternate exterior angles are congruent.

3.15: Two lines are parallel if when cut by a transversal, a pair of same side interior angles are supplementary.

3.16: Two lines are parallel if when cut by a transversal, a pair of same side exterior angles are supplementary.

Chapter 4

Find the 3rd angle and state whether the triangle is acute, right, or obtuse:

4.1:

4.2:

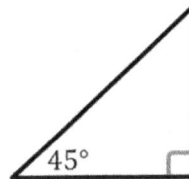

4.3:

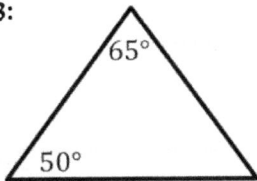

4.4:

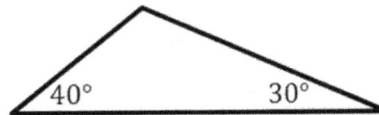

Determine the angle x:

4.5:

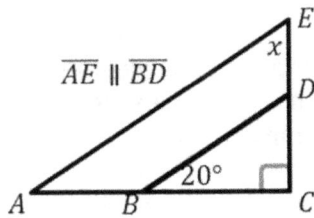

4.6:

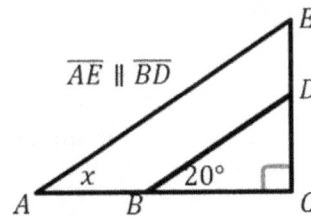

4.7:

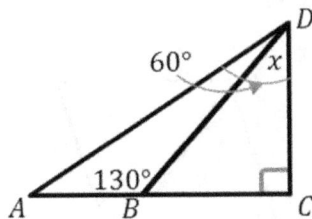

4.8:

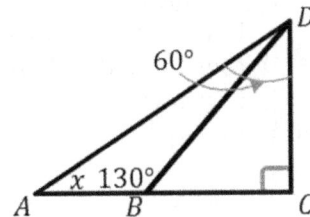

4.9:

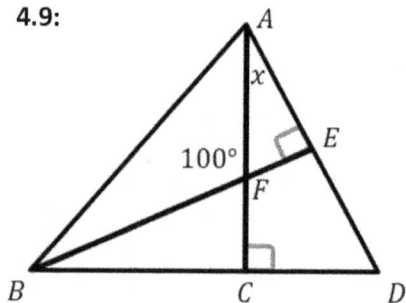

4.10:

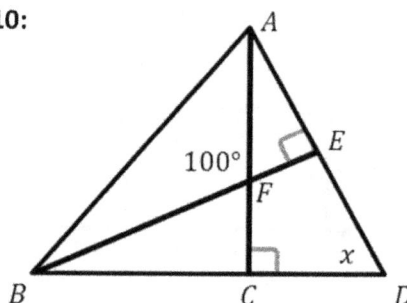

4.11: List the sides from largest to smallest

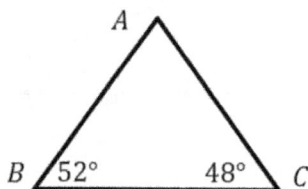

4.12: List the angles from largest to smallest

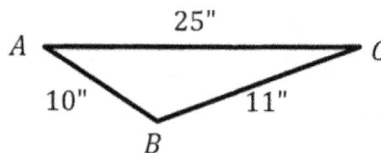

Prove the following:

4.13: A triangle can have no more than one angle that is greater than or equal to 90°.

4.14: For a right triangle, the other two angles are complementary.

4.15: In comparing two triangles, if two of their angles are congruent, then their third angles are congruent.

4.16: $\angle a \cong \angle b$

$\overline{AD} \parallel \overline{BC}$

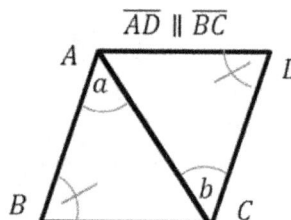

Chapter 5

Determine if the following triangles are congruent to the triangle to the right:

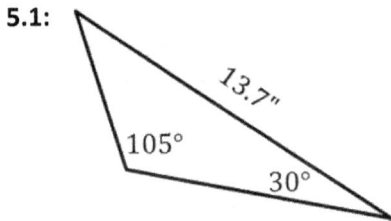

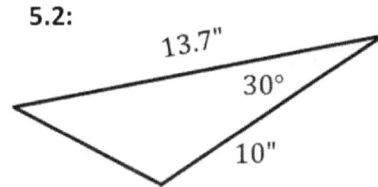

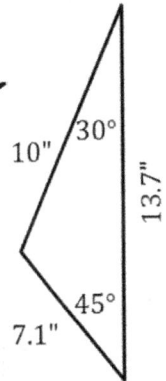

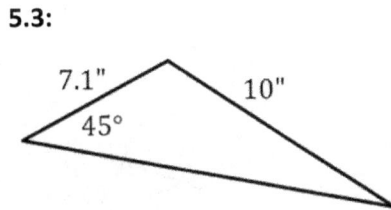

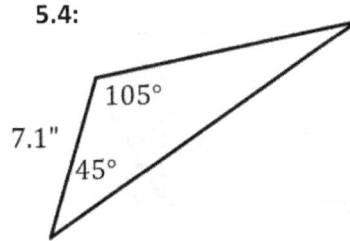

5.1:

13.7"

105°

30°

5.2:

13.7"

30°

10"

10" 30°

13.7"

45°

7.1"

5.3:

7.1" 10"

45°

5.4:

105°

7.1"

45°

Show the indicated triangles are congruent:

5.5: $\triangle CAB \cong \triangle BDC$

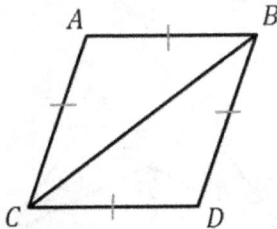

A   B

C   D

5.6: $\triangle ABC \cong \triangle ADC$

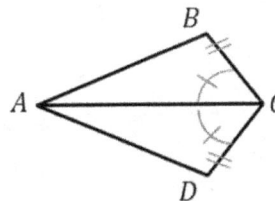

B

A   C

D

5.7: $\triangle ABE \cong \triangle DBC$

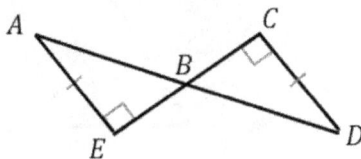

A   C

B

E   D

5.8: $\triangle ABD \cong \triangle CBD$

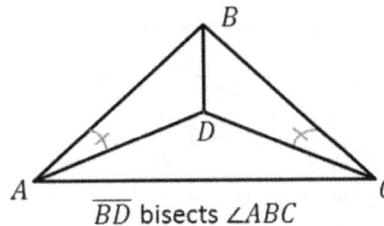

B

D

A   C

$\overline{BD}$ bisects $\angle ABC$

5.9: $\triangle ACD \cong \triangle ECB$

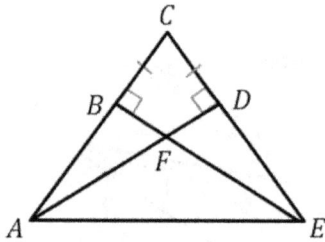

5.10: $\triangle ABE \cong \triangle CBD$

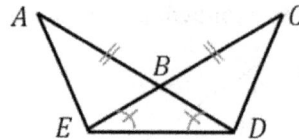

5.11: $\triangle AED \cong \triangle CFB$

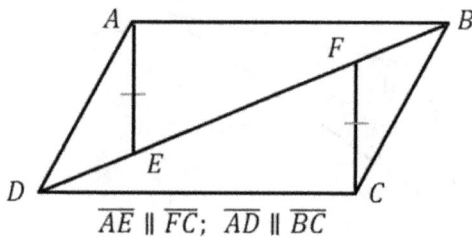

$\overline{AE} \parallel \overline{FC}; \ \overline{AD} \parallel \overline{BC}$

5.12: $\triangle ADE \cong \triangle EBA$

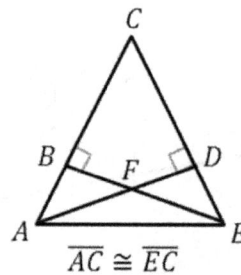

$\overline{AC} \cong \overline{EC}$

Prove the following:

5.13: $\overline{DA} \cong \overline{AB}$

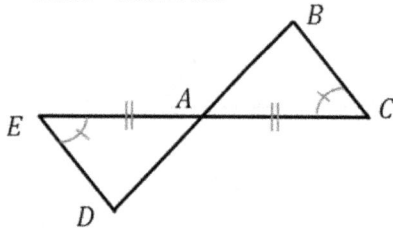

5.14: $\angle B \cong \angle C$

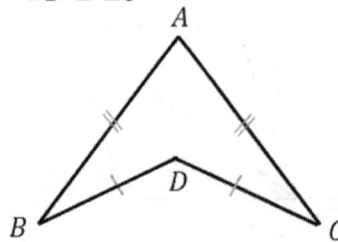

5.15: $\overline{AC} \parallel \overline{BD}$

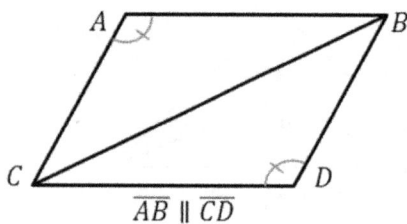

$\overline{AB} \parallel \overline{CD}$

5.16: $\overline{BE} \parallel \overline{CD}$

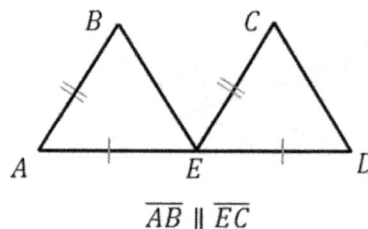

$\overline{AB} \parallel \overline{EC}$

Chapter 6

Show the indicated triangles are similar:

6.1: $\triangle ABE \sim \triangle BAC$

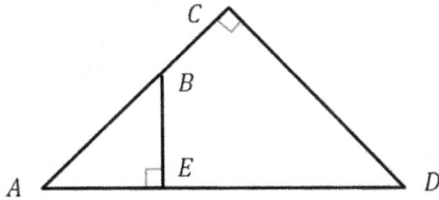

6.2: $\triangle ABC \sim \triangle DEC$

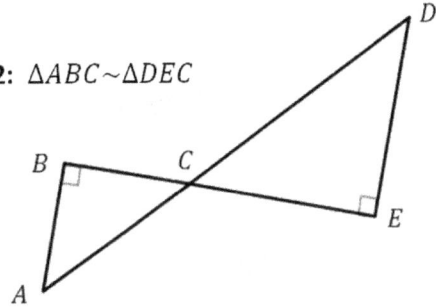

6.3: $\triangle ACE \sim \triangle BCD$

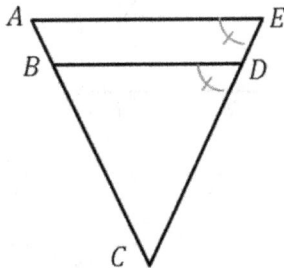

6.4: $\triangle ACB \sim \triangle DCE$

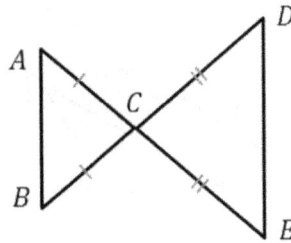

Determine x:

6.5:

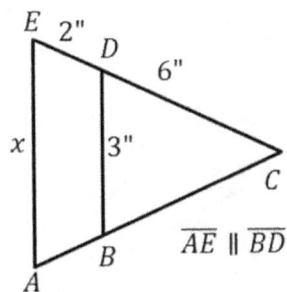

$\overline{AE} \parallel \overline{BD}$

6.6:

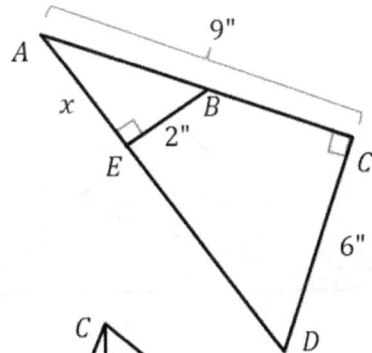

6.7:

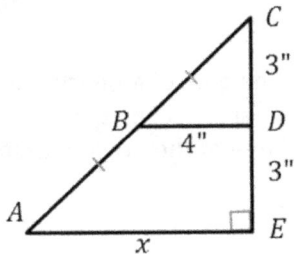

6.8:

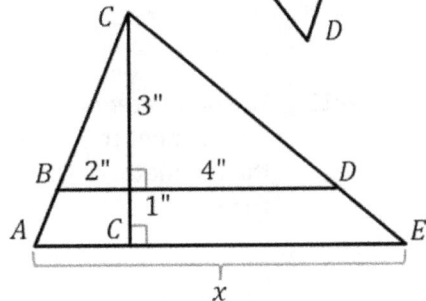

6.9: Find the altitude of $\triangle AOC$

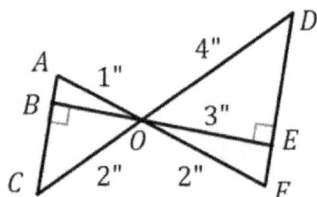

6.10: Find the area of $\triangle BCE$

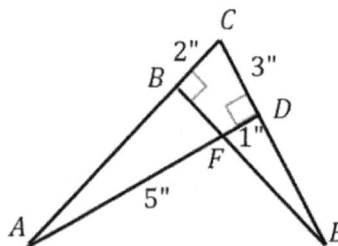

6.11: Find the area of $\triangle ACE$

$BG = 2"$
$FD = 5"$

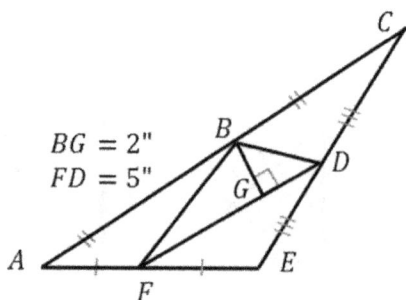

6.12: Find the perimeter of $\triangle BDF$

$9"$

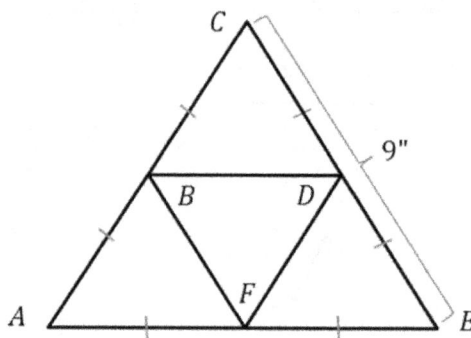

Prove the following:

6.13: $\dfrac{BD}{AE} \cong \dfrac{CD}{CE}$

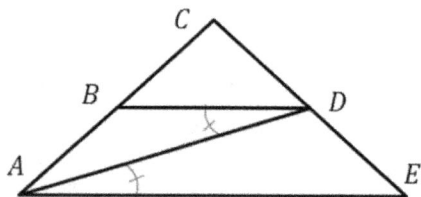

6.14: $\dfrac{BC}{CD} \cong \dfrac{BE}{AD}$

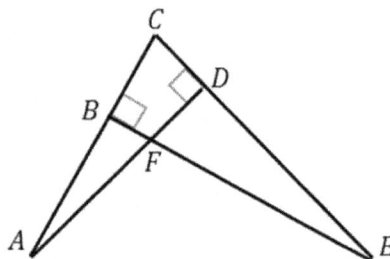

6.15: The perimeter of a midsegment triangle is half the perimeter of the original triangle.

6.16: The area of a midsegment triangle is one quarter the area of the original triangle.

Chapter 7

Determine the measure of all sides:

7.1:

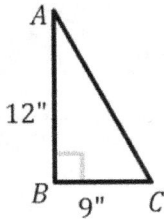

7.2:

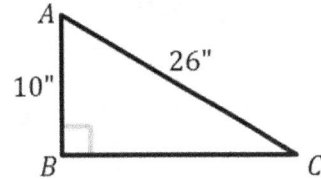

7.3:

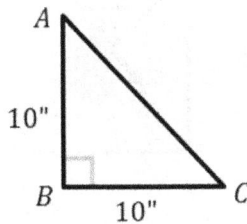

7.4:

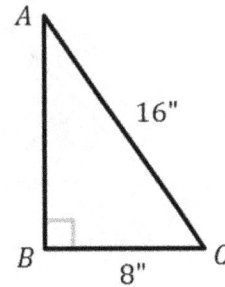

7.5:

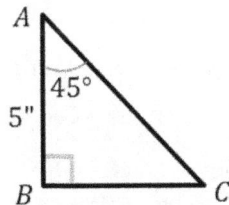

7.6:

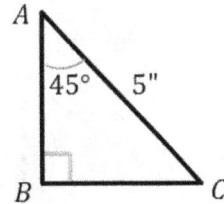

7.7:

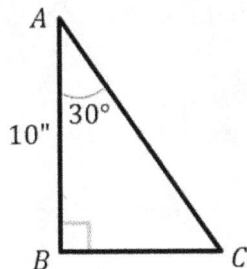

7.8:

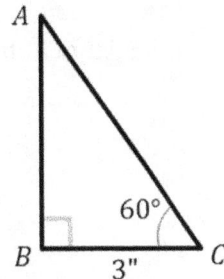

Determine x:

7.9:

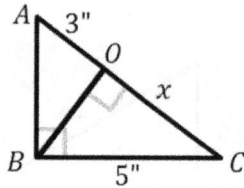

7.10:

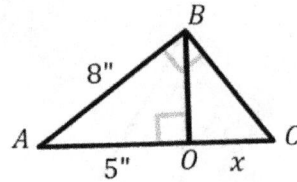

7.11:

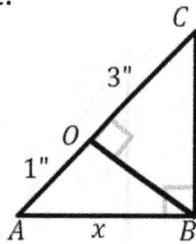

7.12:

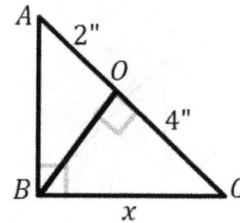

Prove the following:

7.13: $xc = a^2$

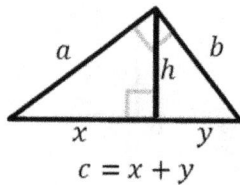

$$c = x + y$$

7.14: $yc = b^2$
using figure from Problem 7.13

7.15: $xy = h^2$
using figure from Problem 7.13

7.16: $Area = \frac{\sqrt{3}}{4}s^2$

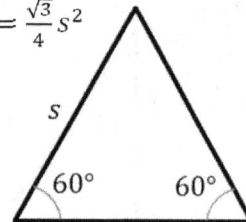

Chapter 8

8.1: Find the internal and external angles of a regular heptagon.

8.2: Find the internal and external angles of a regular decagon.

8.3: Plot the measure of an internal angle versus the number of sides of a regular polygon to see the upper limit on internal angle size.

8.4: If Jasmyne walks 40' due East, turns 20° to the right, walks another 40', turns 20° again to the right, and so on until she reaches her starting point, how many feet did she traverse?

8.5: Find the area of the shaded region (the octagon is regular and the right angle is at the octagon center).

10"

12"

12"

8.6: Find the area of the shaded region (the hexagram is regular).

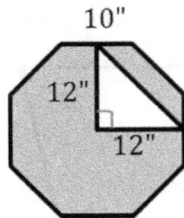

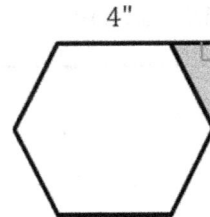

4"

8.7: Find the area of the regular polygon (hint: look at problem 8.6).

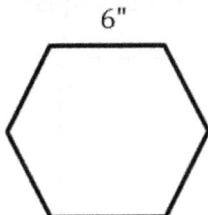

6"

8.8: Find the area of the regular polygon (hint: look at problem 8.6).

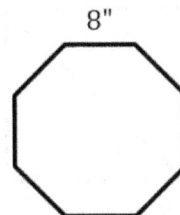

8"

Prove the following:

8.9: *ABCD* is a parallelogram

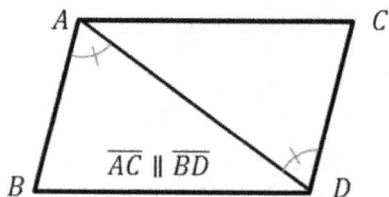

$$\overline{AC} \parallel \overline{BD}$$

8.10: *ABCD* is a parallelogram

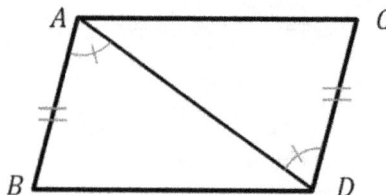

8.11: *ABCD* is a parallelogram

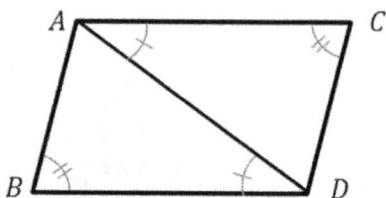

8.12: *ABCD* is a parallelogram

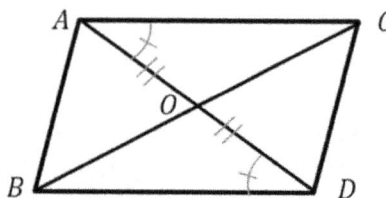

8.13: *ABCD* is a rectangle

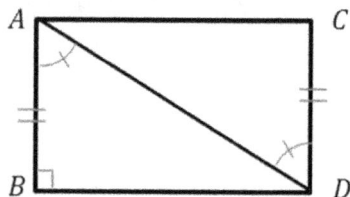

8.14: *ABCD* is a rhombus

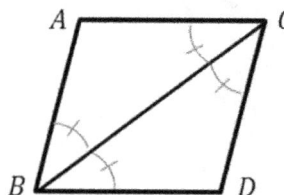

8.15: *ABCD* is a square

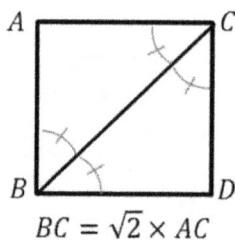

$$BC = \sqrt{2} \times AC$$

8.16: Using the area formula of a parallelogram is $ap/2$, prove the area of a square is s^2.

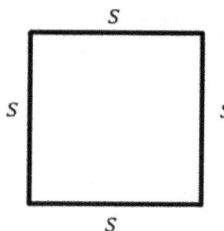

Chapter 9

Refer to the following figure for Problems 9.1-9.6:

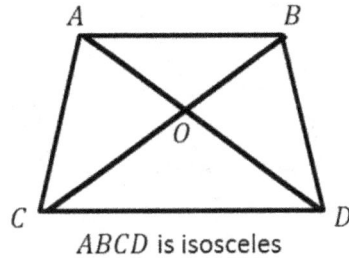

ABCD is isosceles

9.1: Show $\triangle ABC \cong \triangle BAD$

9.2: Show $\triangle AOC \cong \triangle BOD$

9.3: Show $AO = OB$

9.4: Show $\triangle ACD \cong \triangle BDC$

9.5: Show $\triangle COD$ is isosceles

9.6: Show $\triangle BOA \sim \triangle COD$

Solve the following:

9.7: Show $ABCE$ is a parallelogram

9.8: Show $GF = EH$

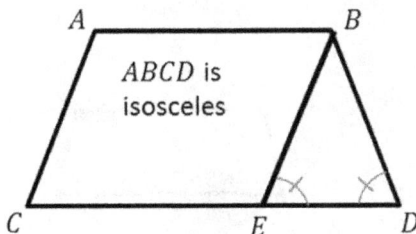

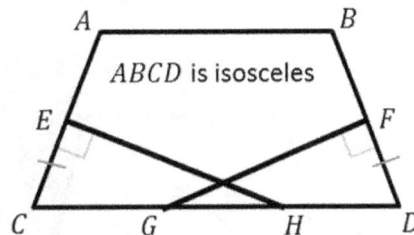

9.9: Find OP

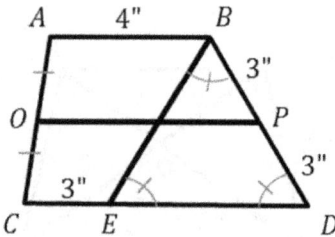

9.10: Find OP

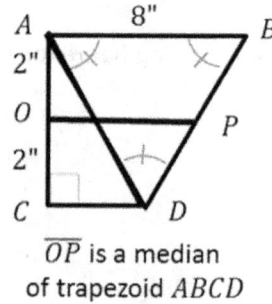

$\overline{OP}$ is a median
of trapezoid $ABCD$

9.11: Find OP

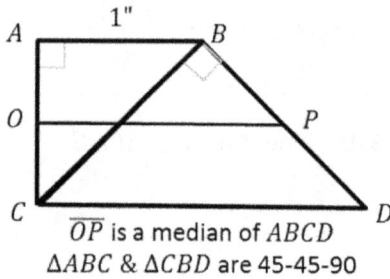

$\overline{OP}$ is a median of $ABCD$
$\triangle ABC$ & $\triangle CBD$ are 45-45-90

9.12: Find BD

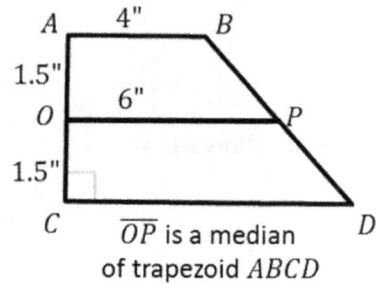

$\overline{OP}$ is a median
of trapezoid $ABCD$

Prove $ABCD$ is an isosceles trapezoid:

9.13:

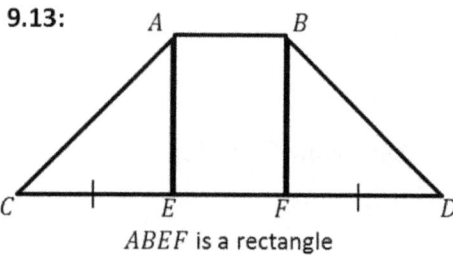

$ABEF$ is a rectangle

9.14:

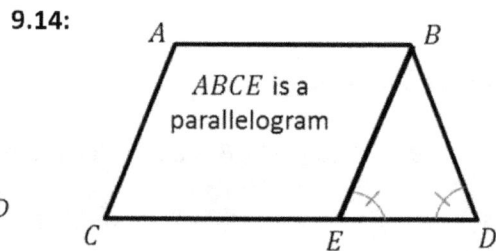

$ABCE$ is a
parallelogram

9.15:

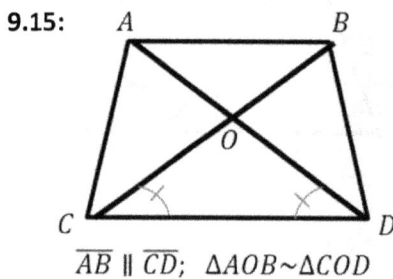

$\overline{AB} \parallel \overline{CD}$; $\triangle AOB \sim \triangle COD$

9.16:

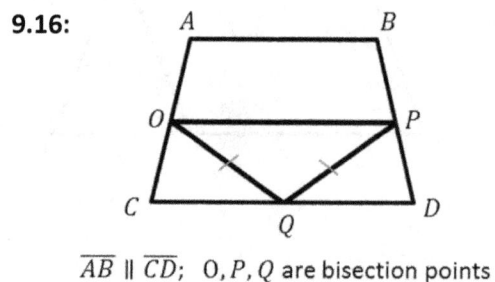

$\overline{AB} \parallel \overline{CD}$; O, P, Q are bisection points

Chapter 10

For Chapter 10 problems, point O or point Q designates the center of a circle:

10.1: Find $m\overline{OQ}$:

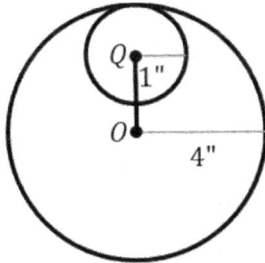

$\odot\,Q$ is tangent to $\odot\,O$.

10.2: Find $m\overline{AB}$:

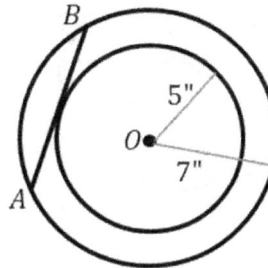

$\overline{AB}$ is tangent to inner circle;
Circles are concentric.

10.3: Find the area of the pentagram.

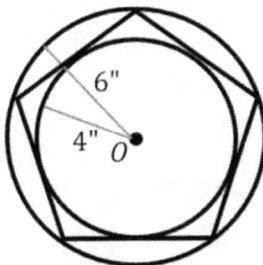

Circles are concentric;
Pentagon is regular,
inscribed, and circumscribed.

10.4: Find the area of the hexagram.

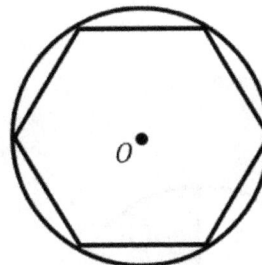

Regular hexagon is inscribed
in circle of diameter 20".

91

Find $m\angle P$:

10.5:

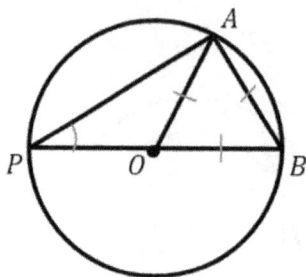

10.6:

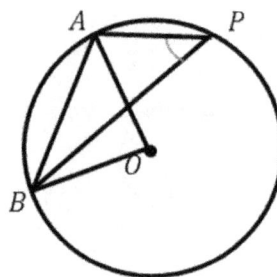

$m\angle BAO = 40°$.

10.7:

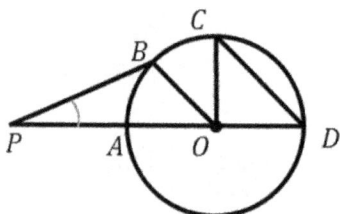

$m\angle OCD = 45°$; $\overline{BO} \parallel \overline{CD}$.

10.8:

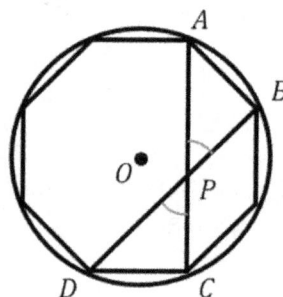

Regular octagon is inscribed.

Find the area indicated:

10.9:

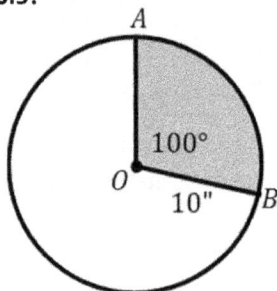

10.10:

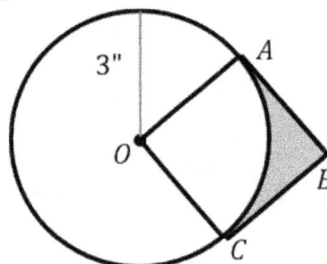

Shape $OABC$ is a square.

10.11:

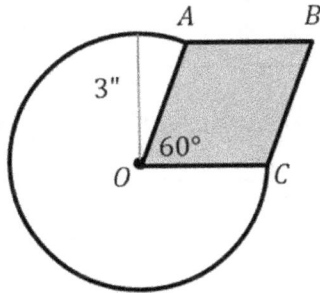

Shape $OABC$ is a rhombus.

10.12: Find the areas of the circles and square in terms of x:

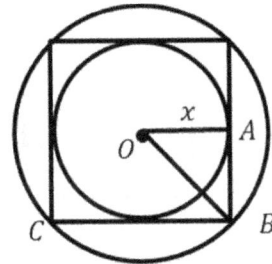

Circles are concentric; Square is Inscribed and circumscribed.

Prove the following:

10.13: $\overline{OP} \cong \overline{PQ}$

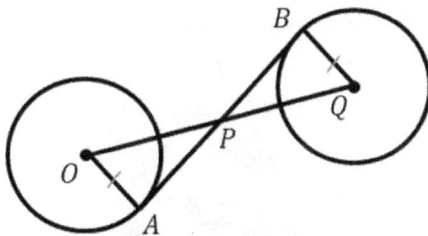

$\overline{AB}$ is tangent to $\odot\, O$ and $\odot\, Q$

10.14: $\overline{OE} \cong \overline{OF}$

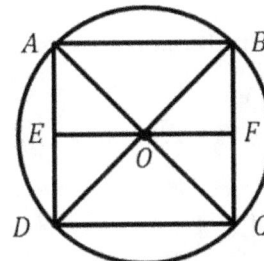

$\overline{AB} \parallel \overline{CD}$.

10.15: The chords between two diameters are congruent.

10.16: The only perpendicular from center circle to a tangent line is at the point of intersection between the tangent and the circle.

Chapter 11

Name all of the symmetries of the following objects:

11.1: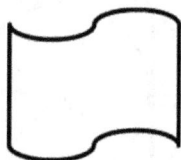

11.2:

11.3:

11.4: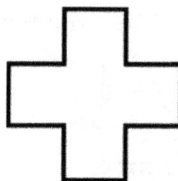

Name the transformations and identify the number of reflections required:

11.5:

11.6:

11.7: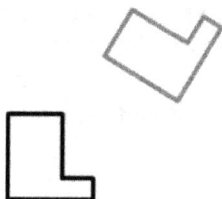

11.8:

Perform the transformations indicated on the following triangle for Problems 11.9-11.16:

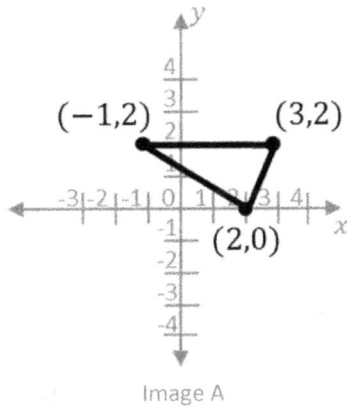

Image A

11.9: $T_{-2,3}(A)$

11.10: $r_{x=1}(A)$

11.11: $r_{y=-1}(A)$

11.12: $r_{(1,1)}(A)$

11.13: $D_2(A)$

11.14: $R_{90°}(A)$

11.15: $T_{1,0} \circ R_{90°}(A)$

11.16: $R_{90°} \circ T_{1,0}(A)$

Chapter 12

Find the loci indicated if they exist:

12.1: 1" ⊥ distance from the line and point as shown below.

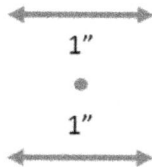

1"

1"

12.2: 1" ⊥ distance from two lines and a point as shown below.

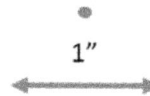

1"

12.3: Equidistant from the two points and 1" from the line

2"
1"   1"

12.4: Equidistant from four pints outlining a 2" square

2"
2"   2"
2"

12.5: Equidistant from the two lines and 1" from the point

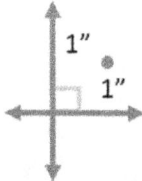

1"
1"

12.6: Equidistant from the two lines and 1" from the circle

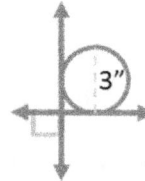

3"

12.7: 1" ⊥ distance from the sides of a 2" square

2"
2"

12.8: 1" ⊥ distance from the sides of a 4" square

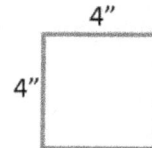

4"
4"

Draw the following:

12.9: Centroid

12.10: Orthocenter

12.11: Circumcenter

12.12: Incenter

Answer the following:

12.13: How far from the top vertex is the centroid of the equilateral triangle given below?

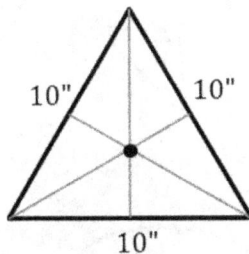

12.14: Is it true that point O is the orthocenter of $\triangle MNO$ and the circumcenter of $\triangle ABC$ if both triangles are 45-45-90.?

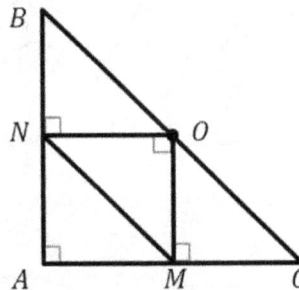

12.15: For an equilateral triangle, which centers share the same location?

12.16: How far apart are the circumcenter and orthocenter of a right isosceles triangle with an 8" hypotenuse?

Chapter 13

13.1: Find the surface area and volume of a right prism with regular octagonal bases as shown.

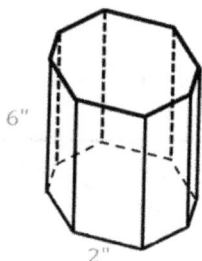

13.2: Find the surface area and volume of a right pyramid with an equilateral triangular base as shown.

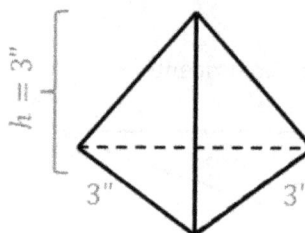

13.3: Compare the volumes of the two solids below to the solid of Problem 13.1.

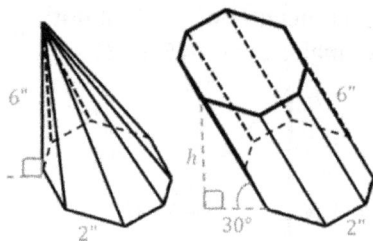

13.4: Compare the volumes of the two solids below to the solid of Problem 13.2.

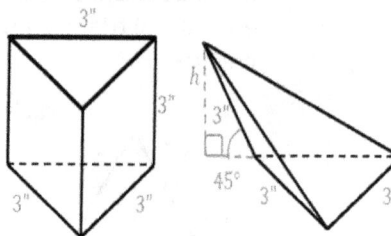

Compare volumes:

13.5:

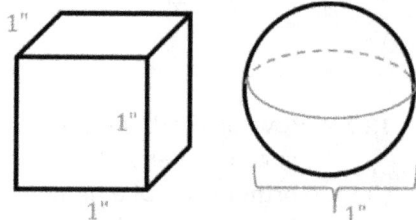

13.6:

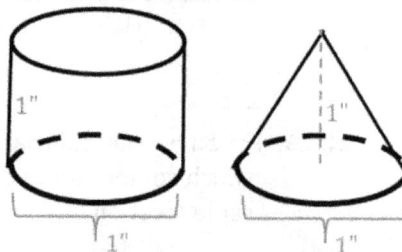

13.7: A cube with side x and a cube with side $2x$.

13.8: A sphere with diameter x and a sphere with diamter $2x$.

Which has the larger volume and which has the largest surface area:

13.9: If the height is doubled or the radius is doubled?

13.10: If the height is doubled or the radius is doubled?

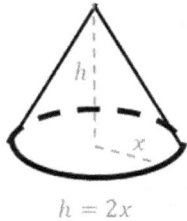

$h = 2x$

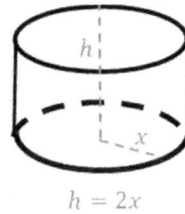

$h = 2x$

Find the volume and surface area of the solids below:

13.11:

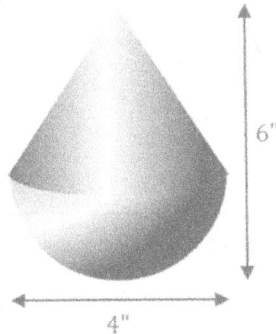

6"

4"

13.12:

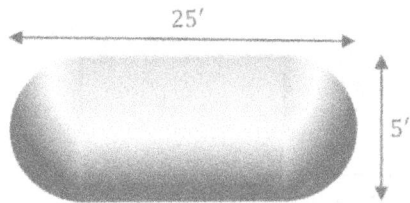

25'

5'

Answer the following:

13.13: Will a sphere with a circumference of 10" fit inside a cube with volume 30 in^3?

13.14: How long is a string of 10 balls laid end to end each with a volume of 9 in^3?

13.15: Which is taller: A sphere or cube each with a volume of 100 in^3?

13.16: If a cone has a 10" diameter base and is 10" high, what is the diameter of a 10" high cylinder with the same volume?

Appendix C: Solutions to Problem Sets

Chapter 1

1.1: $(1,2); (3,4)$	**1.2**: $(-1,-2); (3,4)$
$m = \frac{\Delta y}{\Delta x} = \frac{4-2}{3-1} = \frac{2}{2} = 1;$	$m = \frac{\Delta y}{\Delta x} = \frac{4-(-2)}{3-(-1)} = \frac{6}{4} = \frac{3}{2};$
$midpiont = \left(\frac{1+3}{2}, \frac{2+4}{2}\right) = (2,3);$	$midpiont = \left(\frac{-1+3}{2}, \frac{-2+4}{2}\right) = (1,1);$
$d = \sqrt{(3-1)^2 + (4-2)^2}$	$d = \sqrt{(3-(-1))^2 + (4-(-2))^2}$
$= \sqrt{4+4} = \sqrt{8} = 2\sqrt{2}$	$= \sqrt{16+36} = \sqrt{52} = 2\sqrt{13}$
1.3: $(-5,-3); (-3,0)$	**1.4**: $(4,8); (-5,2)$
$m = \frac{\Delta y}{\Delta x} = \frac{0-(-3)}{-3-(-5)} = \frac{3}{2};$	$m = \frac{\Delta y}{\Delta x} = \frac{2-8}{-5-4} = \frac{-6}{-9} = \frac{2}{3};$
$midpiont = \left(\frac{-5+(-3)}{2}, \frac{-3+0}{2}\right)$	$midpiont = \left(\frac{-5+4}{2}, \frac{2+8}{2}\right) = \left(-\frac{1}{2}, 5\right);$
$= \left(-4, -\frac{3}{2}\right);$	$d = \sqrt{(-5-4)^2 + (2-8)^2}$
$d = \sqrt{(-5-(-3))^2 + (-3-0)^2}$	$= \sqrt{81+36} = \sqrt{117} = 3\sqrt{13}$
$= \sqrt{4+9} = \sqrt{13}$	

1.5:

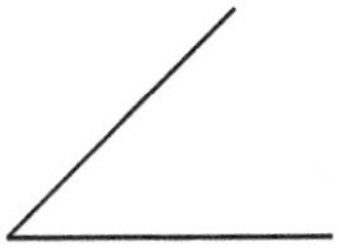

Measurement is 45°

1.6:

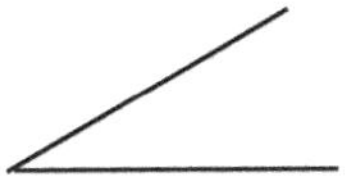

Measurement is 30°

1.7:

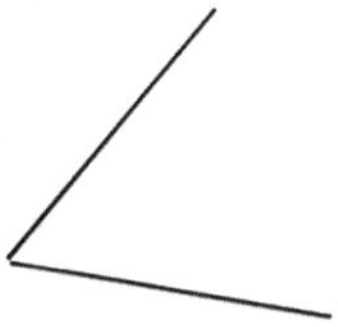

Measurement is 60°

1.8:

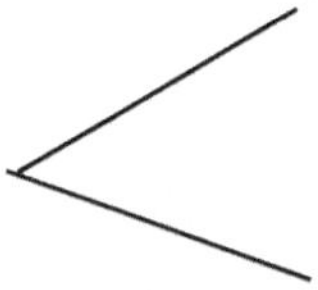

Measurement is 50°

1.9:

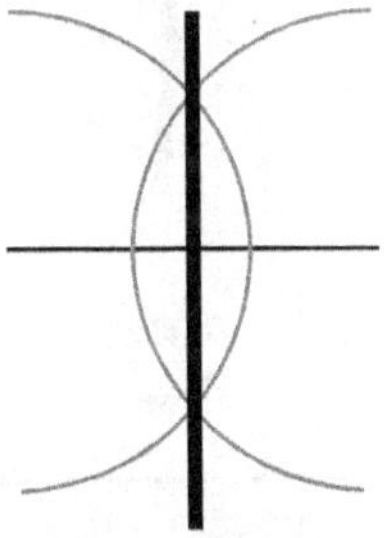

1.10:

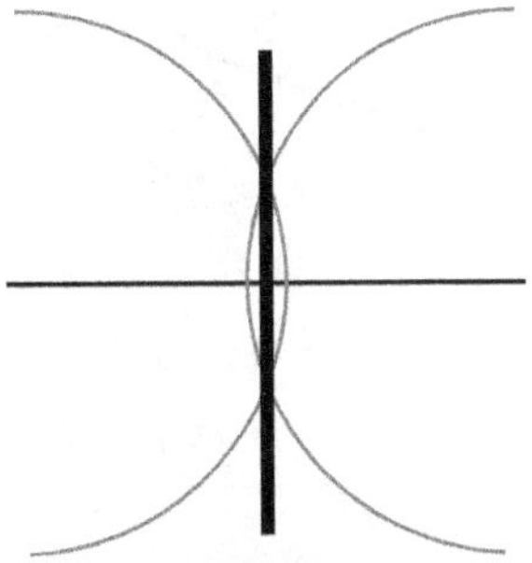

1.11:

1.12:

1.13:

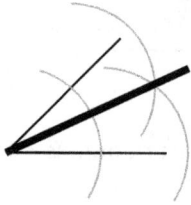

1.14:

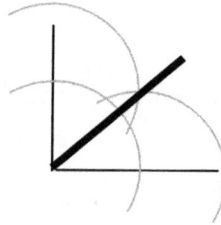

1.15:

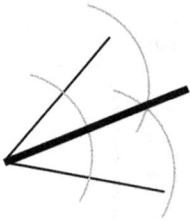

1.16:

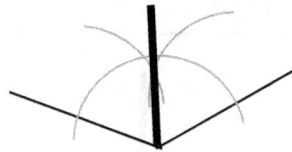

Chapter 2

2.1: You have six pictures in front of you that look similar. You are told two of them are exact matches. How do you determine which two? Rather than evaluate every minute detail of all of the pictures, you find the differences in four of the pictures and end up with two that must be the same. You use indirect reasoning by eliminating all possibilities but one.	**2.2:** Your mother says your room is never clean, but today it is. How do you refute her statement? You can honestly say that there is one example of when your room was clean, therefore it is not true that your room is never clean. You use proof by counterexample.
2.3: You are asked to walk into a perfectly square room and measure all four walls. Why do you only measure one? You use direct reasoning to deduce that since the room is a perfect square, all sides are equal, and therefore you only need to measure one side.	**2.4:** A neighbor tells you that a steady diet of junk food is bad for your weight. You eat 50 candy bars in a week and end up gaining 15 lbs. How did you prove the neighbor right? You used indirect reasoning by assuming the opposite of the premise was true, discovering it wasn't, then realizing the original statement was indeed true.
2.5: $p = q = True$ $(p \wedge \sim q) \vee (p \wedge q)$ $(T \wedge \sim T) \vee (T \wedge T)$ $= (T \wedge F) \vee T$ $= F \vee T$ $= T$	**2.6:** $p = q = False$ $(p \wedge \sim q) \vee (p \wedge q)$ $(F \wedge \sim F) \vee (F \wedge F)$ $= (F \wedge T) \vee F$ $= F \vee F$ $= F$

2.7: $p = q = True$ $(p \lor \sim q) \land (p \lor q)$ $\quad (T \lor \sim T) \land (T \lor T)$ $= (T \lor F) \land T = T \land T = F$	**2.8:** $p = q = False$ $(p \lor \sim q) \land (p \lor q)$ $\quad (F \lor \sim F) \land (F \lor F)$ $= (F \lor T) \land F = T \land F = F$
2.9: $p = \{3 > 2\};$ $\quad q = \{-3 > -2\}$ $(\sim p \land \sim q) \land (p \land q)$ $\quad p = T; \ q = F$ $(\sim T \land \sim F) \land (T \land F)$ $= (F \land T) \land F = F \land F = F$	**2.10:** $p = \{2 + 2 = 5\};$ $\quad q = \{2 \times 4 = 8\}$ $(\sim p \lor \sim q) \lor (p \lor q)$ $\quad p = F; \ q = T$ $(\sim F \lor \sim T) \lor (F \lor T)$ $= (T \lor F) \lor T = T \lor T = T$
2.11: $p = \{AB > AD\};$ $\quad q = \{AD = AC + BD\}$ $(\sim p \land q) \land \sim (p \land q)$ $\quad p = F; \ q = F$ $(\sim F \land F) \land \sim (F \land F)$ $= (T \land F) \land \sim F = F \land T = F$	**2.12:** $p = \{AD > AB\};$ $\quad q = \{AD > AB + CD\}$ $(\sim p \land q) \land \sim (p \land q)$ $\quad p = T; \ q = T$ $(\sim T \land T) \land \sim (T \land T)$ $= (F \land T) \land \sim T = F \land F = F$
2.13: $p = \{AD = AC + CD\};$ $\quad q = \{AD = AB + BD\}$ $\big(q \land (p \land q)\big) \lor \sim p$ $\quad p = T; \ q = T$ $\big(T \land (T \land T)\big) \lor \sim T$ $= (T \land T) \lor F = T \lor F = T$	**2.14:** $p = \{m\angle MLN = m\angle MLO\};$ $\quad q = \{m\angle MLN > m\angle MLO\}$ $\big(q \land (p \land q)\big) \lor \sim p$ $\quad p = F; \ q = F$ $\big(F \land (F \land F)\big) \lor \sim F$ $= (F \land F) \lor T = F \lor T = T$
2.15: $p = \{m\angle MLO > m\angle NLO\};$ $\quad q = \{m\angle NLO > m\angle MLO\}$ $\big(q \land (p \land q)\big) \lor \sim p$ $\quad p = T; \ q = F$ $\big(F \land (T \land F)\big) \lor \sim T$ $= (F \land F) \lor F = F \lor F = F$	**2.16:** $p = \{m\angle MLN < m\angle MLO\};$ $\quad q = \{m\angle NLO < m\angle MLO\}$ $\big(\sim q \land (p \land q)\big) \lor p$ $\quad p = T; \ q = T$ $\big(\sim T \land (T \land T)\big) \lor T$ $= (F \land T) \lor T = F \lor T = T$

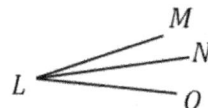

Chapter 3

3.1: 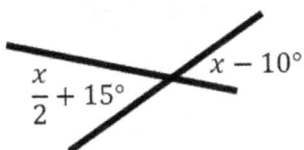 Vertical angles are equal so: $$\frac{x}{2} + 15 = x - 10$$ $$x = 50°$$	**3.2:** 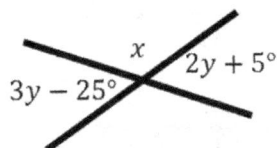 Vertical angles are equal so: $$3y - 25 = 2y + 5$$ $$y = 30°$$ Supplementary angles add to 180°: $$2y + 5 + x = 65 + x = 180$$ $$x = 115°$$
3.3: 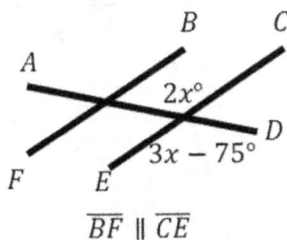 $$\overline{BF} \parallel \overline{CE}$$ Vertical angles are equal so: $$2x = 3x - 75$$ $$x = 75°$$	**3.4:** 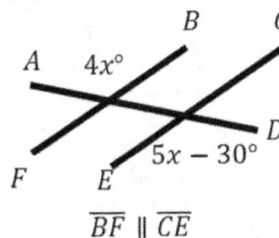 $$\overline{BF} \parallel \overline{CE}$$ Alternating exterior angles are equal so: $$4x = 5x - 30$$ $$x = 30°$$
3.5: 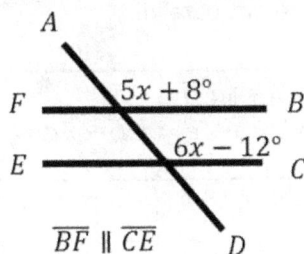 $$\overline{BF} \parallel \overline{CE}$$ Corresponding angles are equal so: $$5x + 8 = 6x - 12$$ $$x = 20°$$	**3.6:** 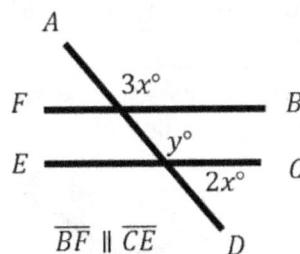 $$\overline{BF} \parallel \overline{CE}$$ $$y = 3x$$ $$y + 2x = 5x = 180$$ $$x = 36°$$

3.7:

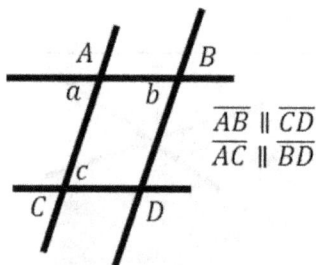

$\overline{AB} \parallel \overline{CD}$
$\overline{AC} \parallel \overline{BD}$

$\angle a$ and $\angle c$ are alternate interior angles of $\overline{AB} \parallel \overline{CD}$, and $\angle a$ and $\angle b$ are corresponding angles of $\overline{AC} \parallel \overline{BD}$, so:

$$\angle a \cong \angle c$$
$$\angle a \cong \angle b$$

And, $\angle b \cong \angle c$ by the transitive property.

3.8:

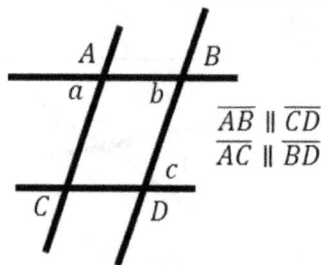

$\overline{AB} \parallel \overline{CD}$
$\overline{AC} \parallel \overline{BD}$

$\angle a$ and $\angle b$ are corresponding angles of $\overline{AC} \parallel \overline{BD}$, and $\angle b$ and $\angle c$ are alternate interior angles of $\overline{AB} \parallel \overline{CD}$, therefore:

$$\angle a \cong \angle b$$
$$\angle c \cong \angle b$$

And, $\angle a \cong \angle c$ by the transitive property.

3.9:

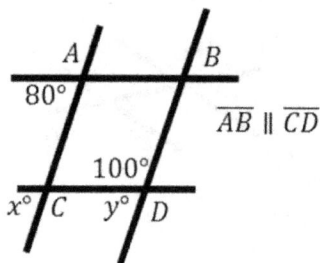

$\overline{AB} \parallel \overline{CD}$

Since $\overline{AB} \parallel \overline{CD}$, $x = 80°$. We also can see by the figure that $y = 180 - 100 = 80°$. If $\overline{AC} \parallel \overline{BD}$, then $x = y = 80°$ which is true.

3.10:

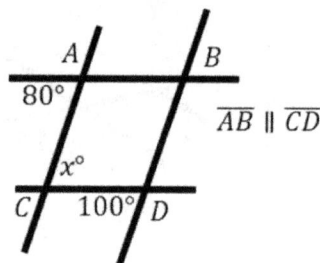

$\overline{AB} \parallel \overline{CD}$

Since $\overline{AB} \parallel \overline{CD}$, $x = 80°$. If $\overline{AC} \parallel \overline{BD}$, then $x = 100°$ which is not true. The lines are not parallel.

3.11: When two parallel lines are cut by a transversal, same side interior angles are supplementary.

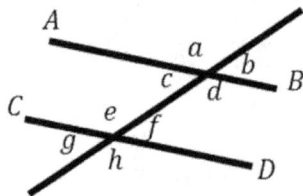

Given: $\overline{AB} \parallel \overline{CD}$
Prove: $\angle c + \angle e = \angle f + \angle d = 180°$

$\angle a + \angle c = 180°$	Supplementary
$\angle a \cong \angle e$	Corresp. angles
$\angle e + \angle c = 180°$	Substitution
$\angle b + \angle d = 180°$	Supplementary
$\angle b \cong \angle f$	Corresp. anlges
$\angle f + \angle d = 180°$	Substitution

3.12: When two parallel lines are cut by a transversal, same side exterior angles are supplementary.

Given: $\overline{AB} \parallel \overline{CD}$
Prove: $\angle a + \angle g = \angle b + \angle h = 180°$

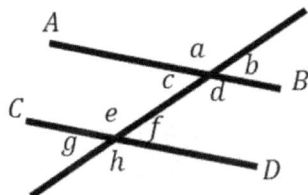

$\angle a + \angle c = 180°$	Supplementary
$\angle c \cong \angle g$	Corresp. angles
$\angle a + \angle g = 180°$	Substitution
$\angle b + \angle d = 180°$	Supplementary
$\angle d \cong \angle h$	Corresp. anlges
$\angle b + \angle h = 180°$	Substitution

3.13: Two lines are parallel if when cut by a transversal, a pair of alternate interior angles are congruent.

Given: $\angle c \cong \angle f$
Prove: $\overline{AB} \parallel \overline{CD}$

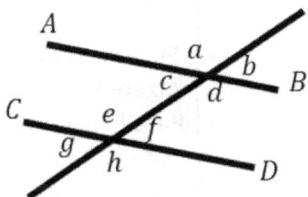

$\angle a + \angle c = 180°$	Supplementary
$\angle e + \angle f = 180°$	Supplementary
$\angle c \cong \angle f$	Given
$180° - \angle a$ $\cong 180° - \angle e$	Substitution
$\angle a \cong \angle e$	Algebra
$\overline{AB} \parallel \overline{CD}$	Parallelism

The last line notes the parallelism postulate on page p 15.

3.14: Two lines are parallel if when cut by a transversal, a pair of alternate exterior angles are congruent.

Given: $\angle a \cong \angle h$
Prove: $\overline{AB} \parallel \overline{CD}$

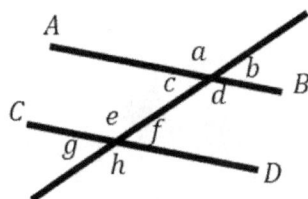

$\angle a \cong \angle h$	Given
$\angle e \cong \angle h$	Vertical angles
$\angle a \cong \angle e$	Transitive
$\overline{AB} \parallel \overline{CD}$	Parallelism

3.15: Two lines are parallel if when cut by a transversal, a pair of same side interior angles are supplementary.

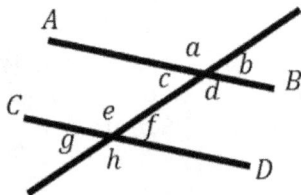

Given: $\angle c + \angle e = 180°$
Prove: $\overline{AB} \parallel \overline{CD}$

$\angle c + \angle e = 180°$	Given
$\angle c + \angle a = 180°$	Supplementary
$180° - \angle a$	
$\cong 180° - \angle e$	Substitution
$\angle a \cong \angle e$	Algebra
$\overline{AB} \parallel \overline{CD}$	Parallelism

3.16: Two lines are parallel if when cut by a transversal, a pair of same side exterior angles are supplementary.

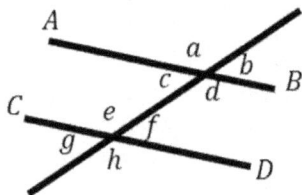

Given: $\angle a + \angle g = 180°$
Prove: $\overline{AB} \parallel \overline{CD}$

$\angle a + \angle g = 180°$	Given
$\angle e + \angle g = 180°$	Supplementary
$180° - \angle a$	
$\cong 180° - \angle e$	Substitution
$\angle a \cong \angle e$	Algebra
$\overline{AB} \parallel \overline{CD}$	Parallelism

Chapter 4

4.1:	4.2:
 $$180° - 30° - 10° = 140°$$ The triangle is obtuse since its largest angle is $> 90°$.	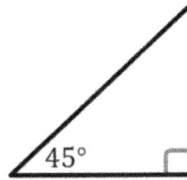 $$180° - 90° - 45° = 45°$$ The triangle is right since its largest angle is equal to $90°$.
4.3:	4.4:
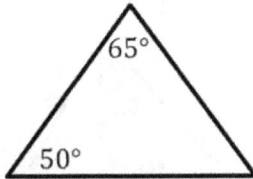 $$180° - 50° - 65° = 65°$$ The triangle is acute since its largest angle is $< 90°$.	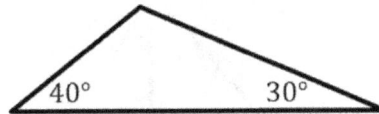 $$180° - 40° - 30° = 110°$$ The triangle is obtuse since its largest angle is $> 90°$.
4.5:	4.6:
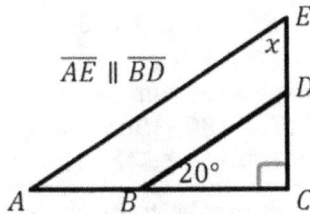 Using triangle angle sum on the smaller triangle: $$m\angle BDC = 180 - 20 - 90 = 70$$ When two parallel lines are cut by a transversal, corresponding angles are equal, so: $$m\angle x = m\angle BDC = 70$$	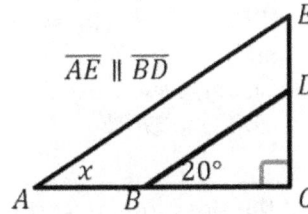 There are two ways to do this. First is to use the triangle angle sum with the results of the prior problem: $$m\angle x = 180 - 70 - 90 = 20$$ The other is to recognize that x is a corresponding angle with the $20°$ angle.

4.7:

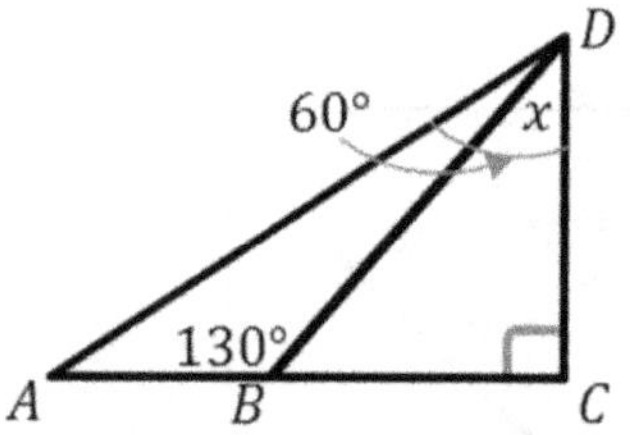

The supplementary angle gives us:
$$m\angle DBC = 180 - 130 = 50$$

Using triangle angle sum:
$$m\angle x = 180 - 50 - 90 = 40$$

4.8:

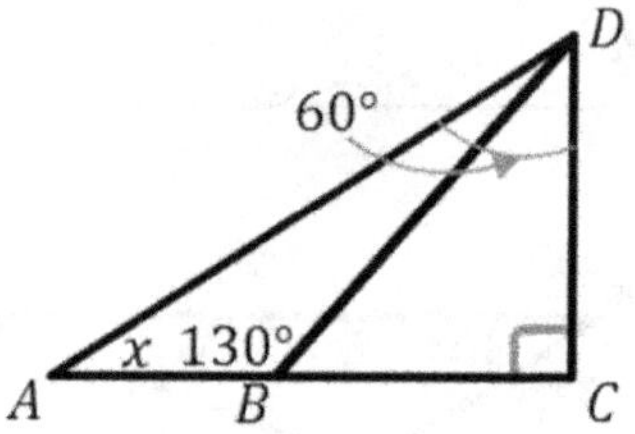

Using triangle angle sum on the larger triangle:
$$m\angle x = 180 - 60 - 90 = 30$$

4.9:

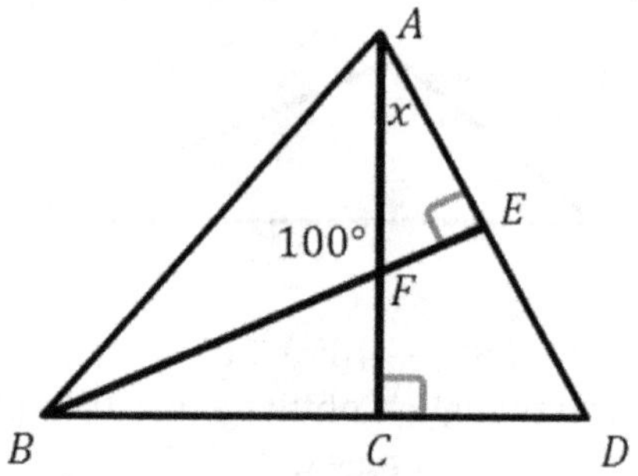

The supplementary angle gives us:
$$m\angle AFE = 180 - 100 = 80$$

Using triangle angle sum:
$$m\angle x = 180 - m\angle AFE - 90 = 10$$

4.10:

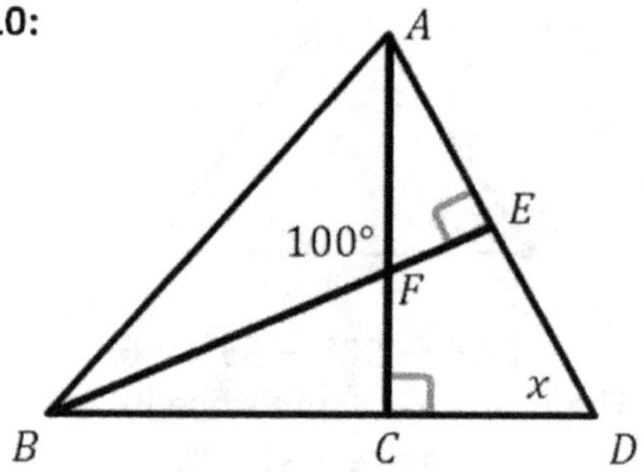

The supplementary angle gives us:
$$m\angle AFE = 180 - 100 = 80$$

Using triangle angle sum:
$$m\angle FAE = 180 - 80 - 90 = 10$$
$$m\angle x = 180 - m\angle FAE - 90 = 80$$

4.11: List the sides from largest to smallest

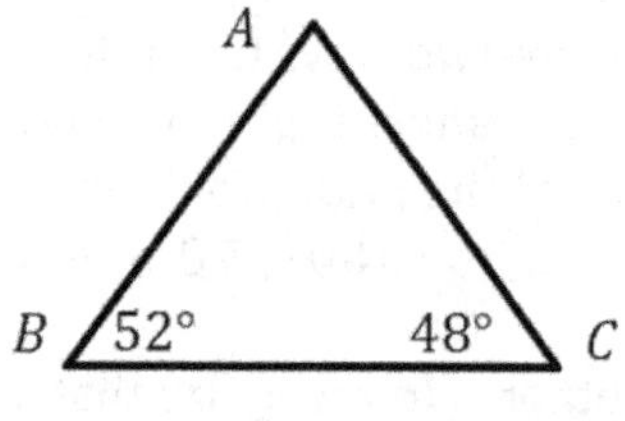

$$m\angle A = 180 - 52 - 48 = 80$$
$$BC > AC > AB$$

4.12: List the angles from largest to smallest

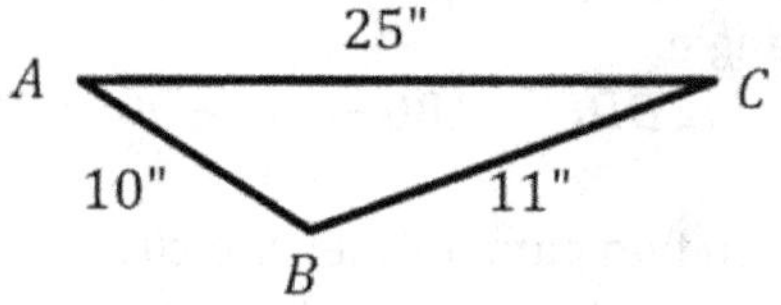

$$m\angle B > m\angle A > m\angle C$$

4.13: A triangle can have no more than one angle that is greater than or equal to 90°.

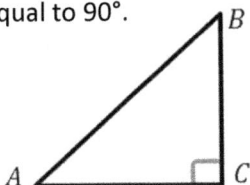

Prove indirectly by assuming there are two right angles.

Given: $\angle B = \angle C = 90°$
Prove: $\angle A$ is nonzero

$\angle B = \angle C = 90°$	Given
$\angle A + \angle B + \angle C = 180°$	Δ angle sum
$\angle A = 180° - 90° - 90°$	Substitution
$\angle A = 0°$	Algebra
$\angle B \neq \angle C = 90°$	Δ's have 3 ∠'s

4.14: For a right triangle, the other two angles are complementary.

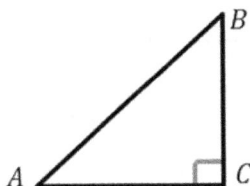

Given: $\angle C = 90°$
Prove: $\angle A + \angle B = 90°$

$\angle C = 90°$	Given
$\angle A + \angle B + \angle C = 180°$	Δ angle sum
$\angle A + \angle B = 180° - 90°$	Substitution
$\angle A + \angle B = 90°$	Algebra

4.15: In comparing two triangles, if two of their angles are congruent, then their third angles are congruent.

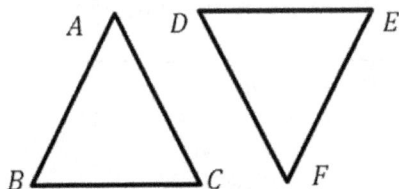

Given: $\angle A \cong \angle F$; $\angle B \cong \angle E$
Prove: $\angle C \cong \angle D$;

$\angle A \cong \angle F$; $\angle B \cong \angle E$	Given
$\angle A + \angle B + \angle C = 180°$	Δ angle sum
$\angle F + \angle E + \angle C = 180°$	Substitution
$\angle F + \angle E + \angle D = 180°$	Δ angle sum
$\angle C - \angle D = 0°$	Subtract equ's
$\angle C \cong \angle D$	Algebra

4.16: $\angle a \cong \angle b$

$\overline{AD} \parallel \overline{BC}$

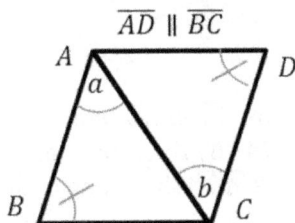

Given: $\overline{AD} \parallel \overline{BC}$; $\angle B \cong \angle D$
Prove: $\angle a \cong \angle b$

$\overline{AD} \parallel \overline{BC}$	Given
$\angle CAD \cong \angle BCA$	Alt. Int. ∠'s & ∥ lines
$\angle B \cong \angle D$	Given
$\angle a \cong \angle b$	Theorem of Prob. 4.15

Chapter 5

5.1: The third angle of the original triangle is: $180° - 30° - 45° = 105°$. The triangles are congruent by AAS.	5.2: The triangles are congruent by SAS.
5.3: Congruency is inconclusive by SSA.	5.4: The triangles are congruent by ASA.
5.5: $\triangle CAB \cong \triangle BDC$ From the figure, $\overline{AB} \cong \overline{CD}$ and $\overline{AC} \cong \overline{BD}$. By the reflexive property, $\overline{CB} \cong \overline{CB}$. The triangles are congruent by SSS.	5.6: $\triangle ABC \cong \triangle ADC$ From the figure, $\overline{BC} \cong \overline{CD}$ and $\angle BCA \cong \angle DCA$. By the reflexive property, $\overline{AC} \cong \overline{AC}$. The triangles are congruent by SAS.

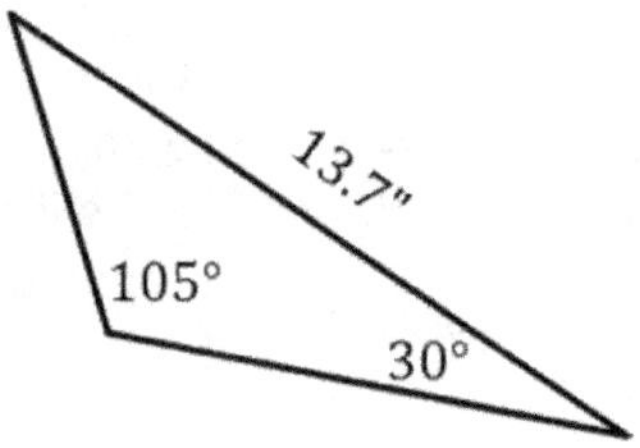

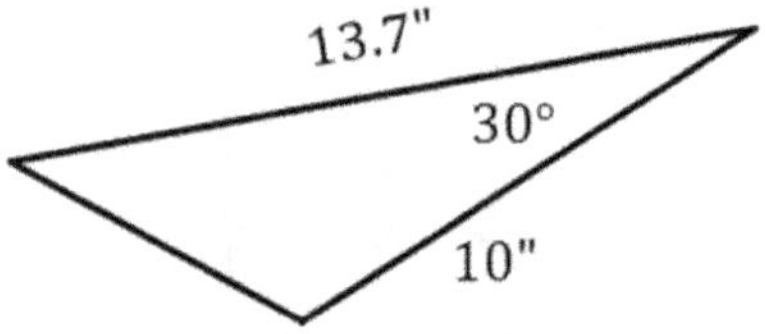

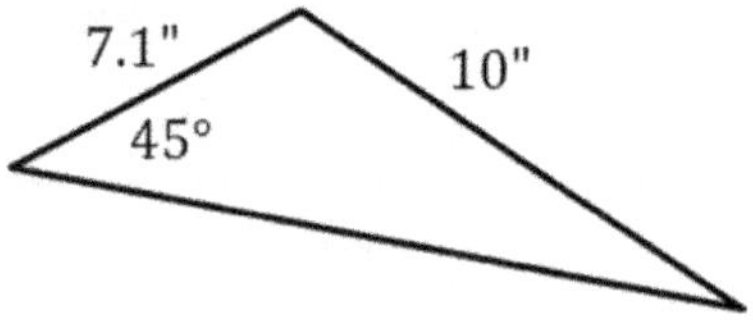

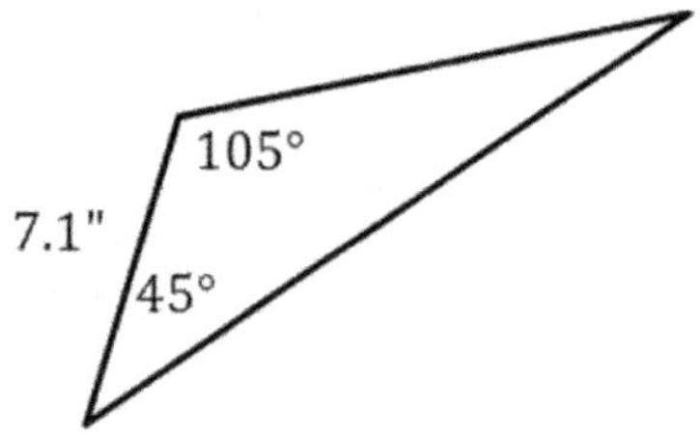

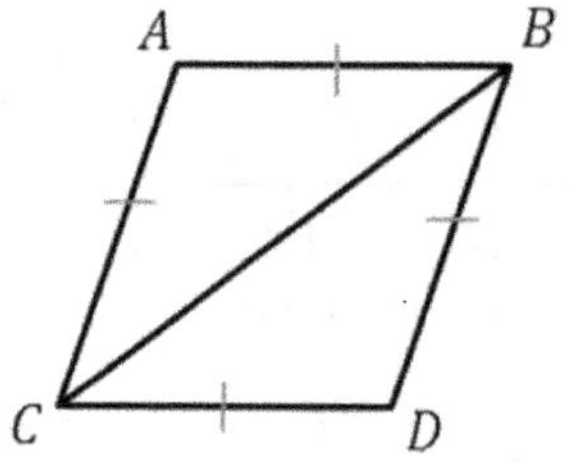

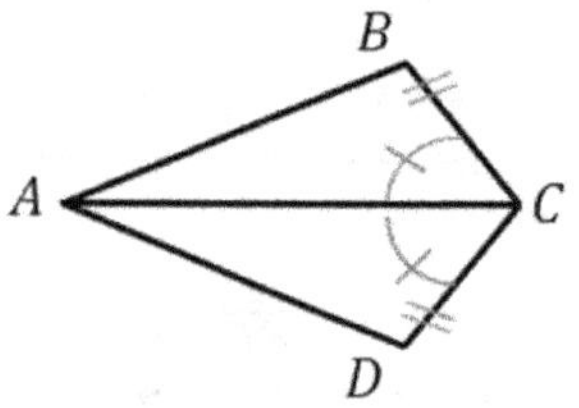

5.7: $\triangle ABE \cong \triangle DBC$

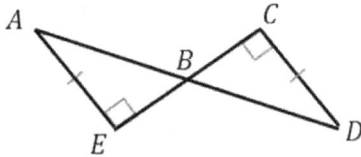

From the figure, $\overline{AE} \cong \overline{CD}$ and $\angle AEB \cong \angle DCB$. By vertical angles, $\angle ABE \cong \angle DBC$. The triangles are congruent by AAS.

5.8: $\triangle ABD \cong \triangle CBD$

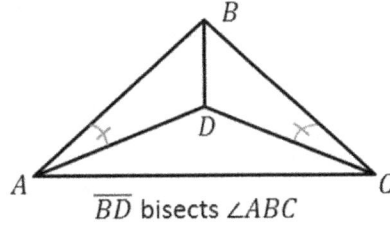

$\overline{BD}$ bisects $\angle ABC$

From the figure, $\angle BAD \cong \angle BCD$. By angle bisectors, $\angle ABD \cong \angle CBD$. By the reflexive property, $\overline{BD} \cong \overline{BD}$. The triangles are congruent by AAS.

5.9: $\triangle ACD \cong \triangle ECB$

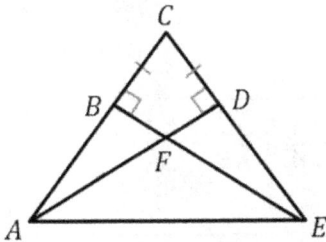

From the figure, $\overline{BC} \cong \overline{CD}$ and $\angle CBE \cong \angle CDA$. By the reflexive property, $\angle C \cong \angle C$. The triangles are congruent by ASA.

5.10: $\triangle ABE \cong \triangle CBD$

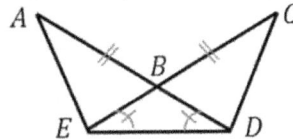

From the figure, $\overline{AB} \cong \overline{BC}$. By vertical angles, $\angle ABE \cong \angle CBD$.

Looking at the bottom $\triangle EBD$, we see $\angle BED \cong \angle BDE$. If two angles are congruent, their opposing sides are congruent, therefore $\overline{EB} \cong \overline{BD}$.

The triangles are congruent by SAS.

5.11: △*AED* ≅ △*CFB*

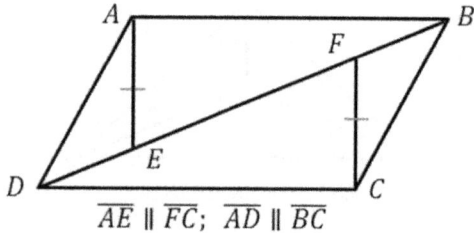

$\overline{AE} \parallel \overline{FC}$; $\overline{AD} \parallel \overline{BC}$

From the figure, $\overline{AE} \cong \overline{FC}$.

Since $\overline{AE} \parallel \overline{FC}$, the alternate interior angles are congruent: $\angle AEF \cong \angle CFE$. They are both supplementary angles, so $\angle AED \cong \angle CFB$.

Since $\overline{AD} \parallel \overline{BC}$, the alternate interior angles are congruent: $\angle ADE \cong \angle CBF$.

The triangles are congruent by AAS.

5.12: △*ADE* ≅ △*EBA*

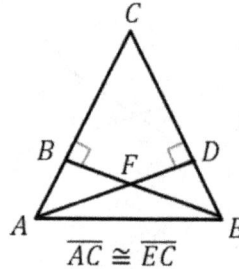

$\overline{AC} \cong \overline{EC}$

From the figure, $\angle ABE \cong \angle EDA$.

Since $\overline{AC} \cong \overline{EC}$, two sides of △*CAE* are congruent meaning their opposing angles are congruent: $\angle BAE \cong \angle DEA$.

By the reflexive property, $\overline{AE} \cong \overline{AE}$.

The triangles are congruent by AAS.

5.13: $\overline{DA} \cong \overline{AB}$

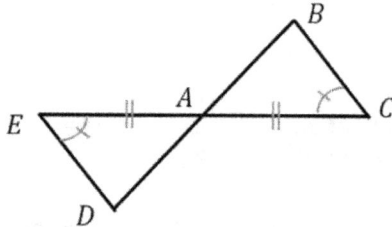

We want to prove congruent triangles so that corresponding elements are equal.

Given: $\overline{EA} \cong \overline{AC}$; $\angle E \cong \angle C$
Prove: $\overline{DA} \cong \overline{AB}$

$\overline{EA} \cong \overline{AC}$	Given
$\angle E \cong \angle C$	Given
$\angle EAD \cong \angle CAB$	Vertical Angles
△*DEA* ≅ △*CBA*	ASA
$\overline{DA} \cong \overline{AB}$	△ Congruency

5.14: $\angle B \cong \angle C$

Given: $\overline{AB} \cong \overline{AC}$; $\overline{BD} \cong \overline{DC}$
Prove: $\angle B \cong \angle C$

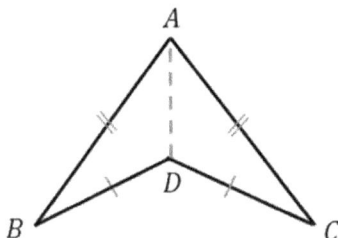

$\overline{AB} \cong \overline{AC}$	Given
$\overline{BD} \cong \overline{DC}$	Given
$\overline{AD} \cong \overline{AD}$	Reflexive
$\triangle ABD \cong \triangle ACD$	SSS
$\angle B \cong \angle C$	$\triangle$ Congruency

Draw $\overline{AD}$ and show $\triangle ABD \cong \triangle ACD$.

5.15: $\overline{AC} \parallel \overline{BD}$

Given: $\overline{AB} \parallel \overline{CD}$; $\angle A \cong \angle D$
Prove: $\overline{AC} \parallel \overline{BD}$

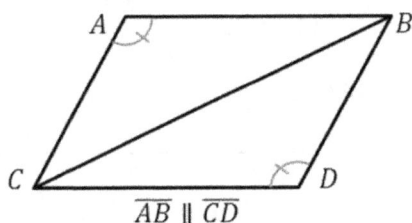

$\overline{AB} \parallel \overline{CD}$

$\angle A \cong \angle D$	Given
$\overline{AB} \parallel \overline{CD}$	Given
$\angle ABC \cong \angle DCB$	Alt. Int. $\angle$'s & $\parallel$ lines
$\overline{CB} \cong \overline{CB}$	Reflexive
$\triangle ABC \cong \triangle DCB$	AAS
$\angle ACB \cong \angle CBD$	$\triangle$ Congruency
$\overline{AC} \parallel \overline{BD}$	Alt. Int. $\angle$'s & $\parallel$ lines

Prove congruent triangles which will give congruent alternate interior angles which then proves parallelism.

5.16: $\overline{BE} \parallel \overline{CD}$

Given: $\overline{AB} \parallel \overline{EC}$; $\overline{AE} \cong \overline{ED}$; $\overline{AB} \cong \overline{EC}$
Prove: $\overline{BE} \parallel \overline{CD}$

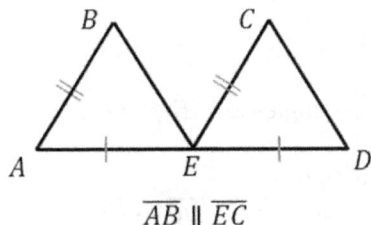

$\overline{AB} \parallel \overline{EC}$

$\overline{AE} \cong \overline{ED}$	Given
$\overline{AB} \cong \overline{EC}$	Given
$\overline{AB} \parallel \overline{EC}$	Given
$\angle A \cong \angle CED$	Corresp. $\angle$'s & $\parallel$ lines
$\triangle ABE \cong \triangle ECD$	SAS
$\angle BEA \cong \angle D$	$\triangle$ Congruency
$\overline{BE} \parallel \overline{CD}$	Corresp. $\angle$'s & $\parallel$ lines

Prove congruent triangles which will give congruent corresponding angles which then proves parallelism.

Chapter 6

6.1: $\triangle ABE \sim \triangle BAC$	**6.2:** $\triangle ABC \sim \triangle DEC$
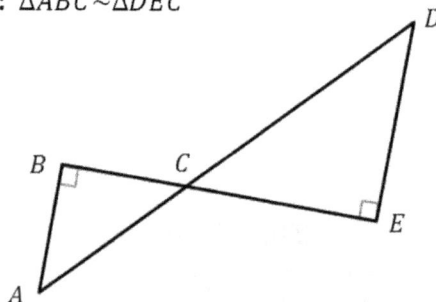	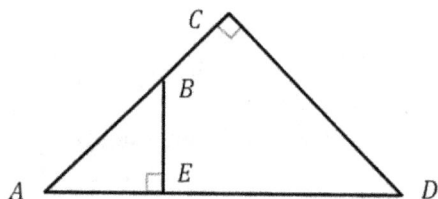
By the figure, $\angle BEA \cong \angle ACD$. By the reflexive property, $\angle A \cong \angle A$. By AA, $\triangle ABE \sim \triangle BAC$	By the figure, $\angle B \cong \angle E$. By vertical angles, $\angle BCA \cong \angle ECD$. By AA, $\triangle ABC \sim \triangle DEC$
6.3: $\triangle ACE \sim \triangle BCD$	**6.4:** $\triangle ACB \sim \triangle DCE$
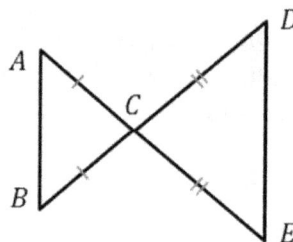	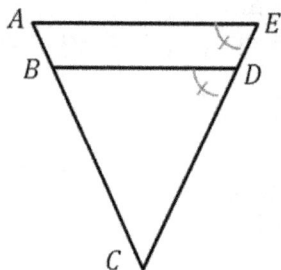
By the figure, $\angle E \cong \angle BDC$. These are corresponding angles which means $\overline{AE} \parallel \overline{BD}$. A line through a triangle parallel to one of its sides produces a similar triangle to the first, so $\triangle ACE \sim \triangle BCD$.	By vertical angles $\angle ACB \cong \angle DCE$. From the figure: $$\overline{AC} \cong \overline{BC}; \ \overline{DC} \cong \overline{EC}.$$ Dividing both equations: $$\frac{AC}{DC} = \frac{BC}{EC}.$$ By SAS, $\triangle ACB \sim \triangle DCE$.

6.5:

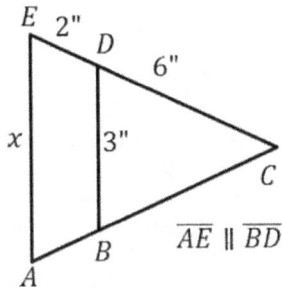

Since $\overline{AE} \parallel \overline{BD}$, we know by corresponding angles $\angle E \cong \angle BDC$. By the reflexive property $\angle C \cong \angle C$. By AA, $\triangle AEC \sim \triangle BDC$. Therefore the sides are in the same proportion:

$$\frac{AE}{BD} = \frac{EC}{DC}$$

$$\frac{x}{3} = \frac{8}{6}$$

$$x = 4"$$

6.6:

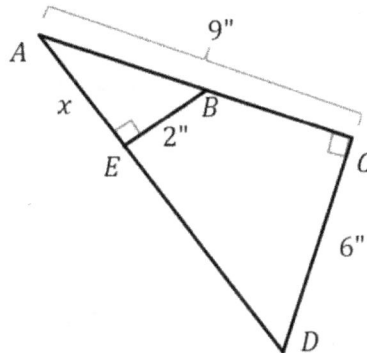

Since $\angle E \cong \angle C$ and $\angle A \cong \angle A$, by AA $\triangle AEB \sim \triangle ADC$. Therefore the sides are in the same proportion:

$$\frac{AE}{AC} = \frac{EB}{DC}$$

$$\frac{x}{9} = \frac{2}{6}$$

$$x = 3"$$

6.7:

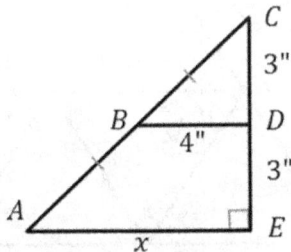

Since $\overline{AB} \cong \overline{BC}$ and $\overline{CD} \cong \overline{DE}$, we know $\overline{BD}$ is a midsegment. Therefore, $\triangle CBD \sim \triangle CAE$ and their sides are in the same proportion:

$$\frac{x}{4} = \frac{6}{3}$$

$$x = 8"$$

6.8:

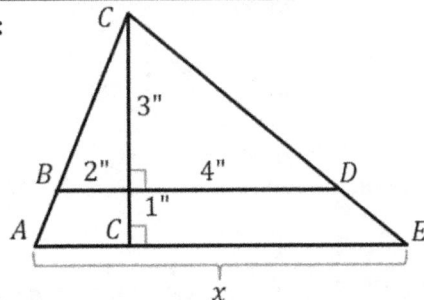

Since the two corresponding right angles are equal, $\overline{AE} \parallel \overline{BD}$ therefore $\triangle ACE \sim \triangle BCD$ and their sides & altitudes are in the same proportion:

$$\frac{x}{6} = \frac{4}{3}$$

$$x = 8"$$

6.9: Find the altitude of $\triangle AOC$

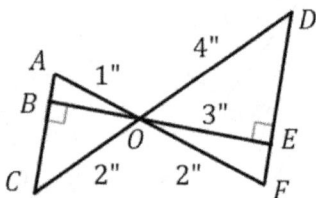

By vertical angles, $\angle AOC \cong \angle DOF$. From the figure we see that

$$\frac{AO}{OF} = \frac{CO}{OD} = 2.$$

By SAS, $\triangle AOC \sim \triangle FOD$. Therefore the sides and altitudes are in the same proportion:

$$BO = 2 \times 3" = 6"$$

6.10: Find the area of $\triangle BCE$

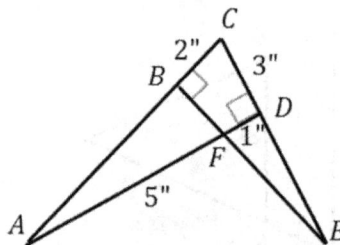

Since $\angle CBE \cong \angle CDA$ and $\angle C \cong \angle C$, by AA, $\triangle ACD \sim \triangle ECB$. Therefore the sides and altitudes are in the same proportion:

$$\frac{EB}{6} = \frac{2}{3} \rightarrow EB = 4"$$

The area of $\triangle BCE$ is then:

$$Area = \frac{1}{2}(EB)(BC) = \frac{1}{2}(4)(2) = 4 \text{ in}^2$$

6.11: Find the area of $\triangle ACE$

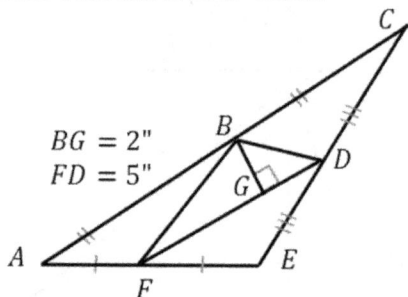

$BG = 2"$
$FD = 5"$

By all of the congruency marks, we can see that $\triangle FBD$ is a midsegment triangle of $\triangle ACE$. The area of $\triangle ACE$ will then be 4 times the area of $\triangle FBD$:

$$Area\ \triangle ACE = 4 \times Area\ \triangle FBD$$

$$= 4\left[\frac{1}{2}(BG)(FD)\right] = 20 \text{ in}^2$$

6.12: Find the perimeter of $\triangle BDF$

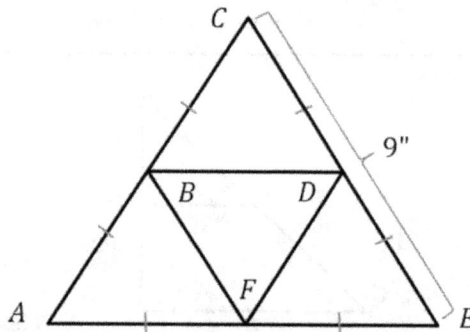

By all of the congruency marks, we can see that $\triangle FBD$ is a midsegment triangle of $\triangle ACE$. The perimeter of $\triangle FBD$ will then be half that of $\triangle ACE$:

$$Perim\ \triangle FBD = \frac{1}{2}(9 + 9 + 9)$$

$$= 13.5 \text{ in}^2$$

6.13: $\frac{BD}{AE} \cong \frac{CD}{CE}$

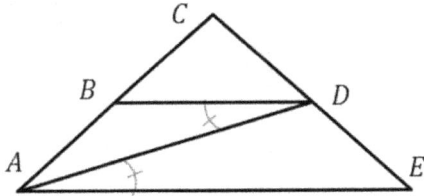

Given: $\angle BDA \cong \angle DAE$

Prove: $\frac{BD}{AE} \cong \frac{CD}{CE}$

$\angle BDA \cong \angle DAE$	Given
$\overline{AE} \parallel \overline{BD}$	Alt. Int. ∠'s & ∥ lines
$\triangle CBD \sim \triangle CAE$	Internal ∥ line in △
$\frac{BD}{AE} \cong \frac{CD}{CE}$	△ Similitude

6.14: $\frac{BC}{CD} \cong \frac{BE}{AD}$

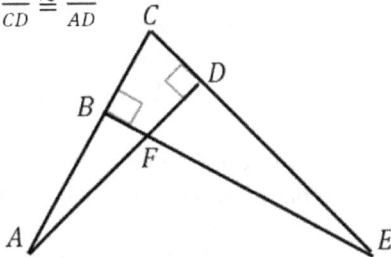

Given: $\angle CDA \cong \angle CBE$

Prove: $\frac{BC}{CD} \cong \frac{BE}{AD}$

$\angle CDA \cong \angle CBE$	Given
$\angle C \cong \angle C$	Reflexive
$\triangle CDA \sim \triangle CBE$	AA
$\frac{BC}{CD} \cong \frac{BE}{AD}$	△ Similitude

6.15: The perimeter of a midsegment triangle is half the perimeter of the original triangle.

6.16: The area of a midsegment triangle is one quarter the area of the original triangle.

We'll do 6.15 & 6.16 together:

Given:
  $\triangle BFD$ is a midsegment △ of $\triangle ACE$
Prove:
  $BD + DF + FB = \frac{1}{2}(AE + AC + EC)$;
  $\frac{1}{2}(BG)(DF) = \frac{1}{4}\left[\frac{1}{2}(EH)(AC)\right]$

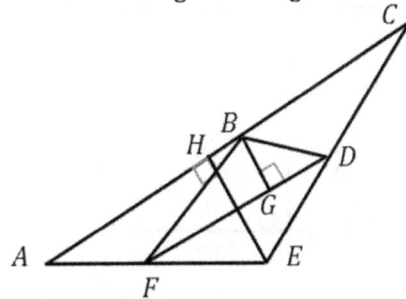

$\triangle BFD$ is a midsegment △ of $\triangle ACE$	Given
$BD = \frac{1}{2}AE$; $DF = \frac{1}{2}AC$; $FB = \frac{1}{2}EC$	Midsegments are ½ length of opp. side
$BD + DF + FB = \frac{1}{2}(AE + AC + EC)$;	Algebra
$\triangle ACE \sim \triangle DFB$	SSS (from 2nd line of proof)
$BG = \frac{1}{2}EH$	Alt. of sim. △'s are in same prop. as sides
$\frac{1}{2}(BG)(DF) = \frac{1}{2}\left[\frac{EH}{2}\left(\frac{AC}{2}\right)\right] = \frac{1}{4}\left[\frac{(EH)(AC)}{2}\right]$	Substitution & Algebra

Chapter 7

7.1: 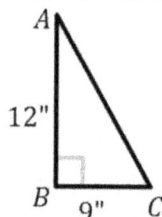	**7.2:** 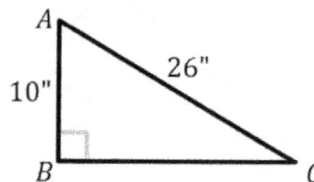
We may notice that the triangle is similar to a 3:4:5 right triangle with a scale factor of 3. Therefore, the hypotenuse is equal to $5 \times 3 = 15"$. If we hadn't noticed this similarity, we could have used Pythagorean's theorem: $$AC^2 = AB^2 + BC^2$$ $$AC = \sqrt{12^2 + 9^2} = \sqrt{225} = 15"$$	We may notice that the triangle is similar to a 5,12,13 right triangle with a scale factor of 2. Therefore, $$BC = 12 \times 2 = 24".$$ If we hadn't noticed this similarity, we could have used Pythagorean's theorem: $$AC^2 = AB^2 + BC^2$$ $$BC = \sqrt{26^2 - 10^2} = \sqrt{576} = 24"$$
7.3: 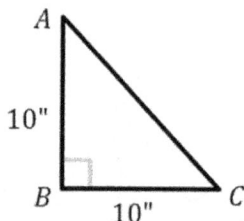	**7.4:** 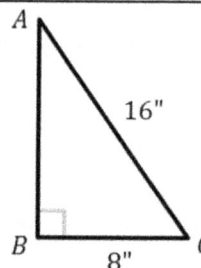
With two equal sides, the triangle is a 45-45-90 with a side ratio of $1:1:\sqrt{2}$. Then, $AC = 10 \times \sqrt{2} = 14.1"$. Alternatively, we can use Pythagorean's theorem: $$AC = \sqrt{10^2 + 10^2} = \sqrt{200} = 14.1"$$	We may notice this is a 30-60-90 triangle with a side ratio of $1:\sqrt{3}:2$. Then, $AB = 8 \times \sqrt{3} = 13.9"$. Alternatively, we can use Pythagorean's theorem: $$AB = \sqrt{16^2 - 8^2} = \sqrt{192} = 13.9"$$

7.5:

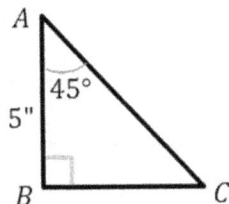

Here we have to notice that the triangle is a 45-45-90 with a side ratio of $1:1:\sqrt{2}$. Then,

$$\frac{AB}{1} = \frac{BC}{1} = \frac{AC}{\sqrt{2}}$$

$$BC = 5 \times 1 = 5"$$

$$AC = 5 \times \sqrt{2} = 7.1"$$

7.6:

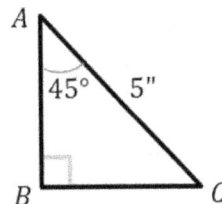

Here we have to notice that the triangle is a 45-45-90 with a side ratio of $1:1:\sqrt{2}$. Then,

$$\frac{AB}{1} = \frac{BC}{1} = \frac{AC}{\sqrt{2}}$$

$$AB = BC = 5 \div \sqrt{2} = 3.5"$$

7.7:

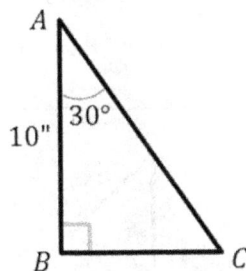

Here we have to notice that the triangle is a 30-60-90 with a side ratio of $1:\sqrt{3}:2$. The smaller side is opposite the smaller angle, so:

$$\frac{AB}{\sqrt{3}} = \frac{BC}{1} = \frac{AC}{2}$$

$$BC = 10 \div \sqrt{3} = 5.8"$$

$$AC = 10 \times 2 \div \sqrt{3} = 11.5"$$

7.8:

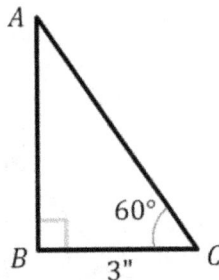

Here we have to notice that the triangle is a 30-60-90 with a side ratio of $1:\sqrt{3}:2$. The smaller side is opposite the smaller angle, so:

$$\frac{AB}{\sqrt{3}} = \frac{BC}{1} = \frac{AC}{2}$$

$$AB = 3 \times \sqrt{3} = 5.2"$$

$$AC = 3 \times 2 = 6"$$

7.9:

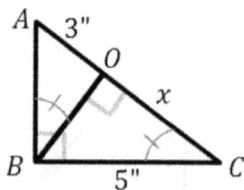

We start by marking a pair of congruent angles as shown. Using similarity with an altitude in a right triangle ($\triangle ABC \sim \triangle BOC$):

$$\frac{AB}{OB} = \frac{BC}{OC} = \frac{AC}{BC} = \frac{5}{x} = \frac{3+x}{5}$$

$$x^2 + 3x - 25 = 0$$

$$x = \frac{-3 \pm \sqrt{9 - 4(-25)}}{2} = \{-6.7, 3.7\}$$

$$x = 3.7"$$

7.10:

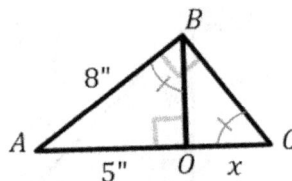

We start by marking a pair of congruent angles as shown. Using similarity with an altitude in a right triangle ($\triangle ABC \sim \triangle AOB$):

$$\frac{AB}{OA} = \frac{BC}{OB} = \frac{AC}{AB} = \frac{8}{5} = \frac{5+x}{8}$$

$$5x + 25 - 64 = 0$$

$$x = 39/5 = 7.8"$$

7.11:

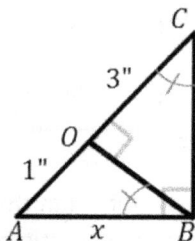

Using similarity ($\triangle ABC \sim \triangle AOB$):

$$\frac{AB}{OA} = \frac{BC}{OB} = \frac{AC}{AB} = \frac{x}{1} = \frac{4}{x}$$

$$x^2 = 4$$

$$x = \{-2, 2\}; \ x = 2"$$

7.12:

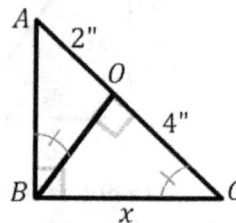

Using similarity ($\triangle ABC \sim \triangle BOC$):

$$\frac{AB}{OB} = \frac{BC}{OC} = \frac{AC}{BC} = \frac{x}{4} = \frac{6}{x}$$

$$x^2 = 24$$

$$x = \{-4.9, 4.9\}; \ x = 4.9"$$

7.13: $xc = a^2$

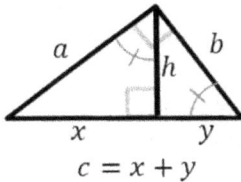

$c = x + y$

Given: $\triangle abc$ is a right triangle;
$\qquad$ h is an altitude of $\triangle abc$
Prove: $xc = a^2$

$\triangle abc$ is right	Given
h is altitude of $\triangle abc$	Given
$\triangle abc \sim \triangle xha$	Right $\triangle$ w/ Alt.
$\dfrac{a}{x} = \dfrac{b}{h} = \dfrac{c}{a}$	$\triangle$ Similitude
$xc = a^2$	Algebra

7.14: $yc = b^2$

Given: $\triangle abc$ is a right triangle;
$\qquad$ h is an altitude of $\triangle abc$
Prove: $yc = b^2$

$\triangle abc$ is right	Given
h is altitude of $\triangle abc$	Given
$\triangle abc \sim \triangle hyb$	Right $\triangle$ w/ Alt.
$\dfrac{a}{h} = \dfrac{b}{y} = \dfrac{c}{b}$	$\triangle$ Similitude
$yc = b^2$	Algebra

7.15: $xy = h^2$

Given: $\triangle abc$ is a right triangle;
$\qquad$ h is an altitude of $\triangle abc$
Prove: $xy = h^2$

$\triangle abc$ is right	Given
h is altitude of $\triangle abc$	Given
$\triangle hyb \sim \triangle xha$	Right $\triangle$ w/ Alt.
$\dfrac{h}{x} = \dfrac{y}{h} = \dfrac{b}{a}$	$\triangle$ Similitude
$xy = h^2$	Algebra

7.16: $Area = \dfrac{\sqrt{3}}{4} s^2$

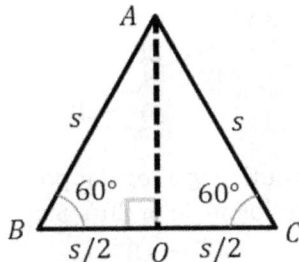

We note that the figure is an equilateral triangle, so each side is the same. Further, we drop an altitude to make two 30-60-90 triangles. We'll sum the area of the two triangles.

Given: Figure to the left
Prove: $Area = \dfrac{\sqrt{3}}{4} s^2$

$\angle B = \angle C = 60°$	Given
$\angle A = 60°$	$\triangle$ Angle Sum
$AB = BC = AC = s$	Equilateral $\triangle$
$BO = OC = s/2$	$\perp$ Bisector equist. from endpoints
$\triangle BAO \cong \triangle CAO$	ASA
$AO = \dfrac{\sqrt{3}s}{2}$	30-60-90 $\triangle$

From the above, we can state:

$$Area\ \triangle BAO = \frac{1}{2}\left(\frac{s}{2}\right)\left(\frac{\sqrt{3}s}{2}\right) = \frac{\sqrt{3}s^2}{8}$$

$$Area\ \triangle ABC = 2 \times Area\ \triangle BAO = \frac{\sqrt{3}}{4} s^2$$

Chapter 8

8.1: Find the internal and external angles of a regular heptagon.

A regular heptagon has 7 sides. The sum of interior angles is then:
$$(7 - 2) \times 180° = 900°$$

Each interior angle is the same, so:
$$\frac{900°}{7} = 128.6°$$

Exterior angles add to 360°, so each is:
$$\frac{360°}{7} = 51.4°$$

8.2: Find the internal and external angles of a regular decagon.

A regular decagon has 10 sides. The sum of interior angles is then:
$$(10 - 2) \times 180° = 1440°$$

Each interior angle is the same, so:
$$\frac{1440°}{10} = 144°$$

Exterior angles add to 360°, so each is:
$$\frac{360°}{10} = 36°$$

8.3: Plot the measure of an internal angle versus the number of sides of a regular polygon to see the upper limit on internal angle size.

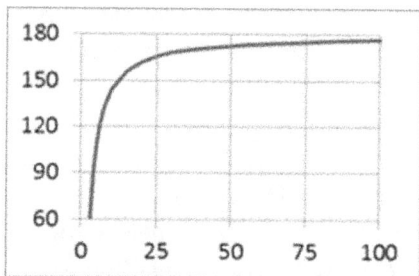

We use the formula :
$$(n - 2) \times 180°/n$$
for the interior angle with n being the number of sides. Starting with $n = 3$, a triangle, we increase the number of sides until the graph rolls off. As might be expected, the maximum interior angle must be less than 180°, otherwise two adjacent sides would form a straight line.

8.4: If Jasmyne walks 40' due East, turns 20° to the right, walks another 40', turns 20° again to the right, and so on until she reaches her starting point, how many feet did she traverse?

The first few turns are shown below.

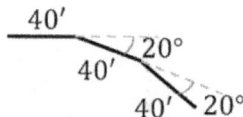

The walk is outlining a regular polygon with an exterior angle of 20°. Since
$$\frac{360}{20} = 18$$

the polygon has 18 sides. The perimeter of the polygon is then
$$18 \times 40' = 720'$$

or just over 1/10th of a mile.

8.5: Find the area of the shaded region (the octagon is regular and the right angle is at the octagon center).

10"

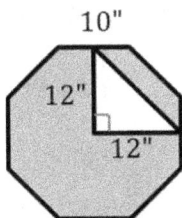

The area of the octagon is:
$$\frac{ap}{2} = \frac{12 \times (8 \times 10)}{2} = 480 \text{ in}^2.$$

The area of the triangle is:
$$\frac{bh}{2} = \frac{144}{2} = 72 \text{ in}^2.$$

The area of the shaded region is then:
$$480 - 72 = 408 \text{ in}^2.$$

8.6: Find the area of the shaded region (the hexagram is regular).

4"

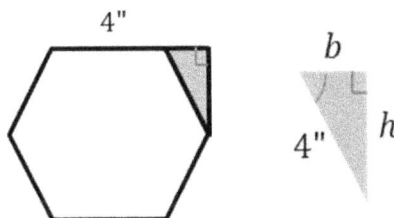

We have a right triangle with 4" hypotenuse. The upper left angle is an exterior angle of the regular hexagram. It's measure must be $\frac{360°}{6} = 60°$. We then have a 30-60-90 triangle so:
$$\frac{b}{1} = \frac{h}{\sqrt{3}} = \frac{4}{2} \rightarrow b = 2; \ h = 2\sqrt{3};$$

The area is then:
$$\frac{bh}{2} = \frac{4\sqrt{3}}{2} = 2\sqrt{3} \text{ in}^2.$$

8.7: Find the area of the regular polygon (hint: look at problem 8.6).

6"

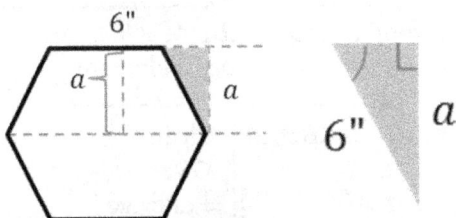

The perimeter is simply 6" × 6 = 36". To find the apothem, we draw a right triangle whose upper left angle is the exterior angle of the hexagram, or $\frac{360°}{6} = 60°$. With a 30-60-90 triangle:
$$\frac{a}{\sqrt{3}} = \frac{6}{2} \rightarrow a = 3\sqrt{3};$$

The area is then:
$$\frac{ap}{2} = \frac{36 \times 3\sqrt{3}}{2} = 54\sqrt{3} \text{ in}^2.$$

8.8: Find the area of the regular polygon (hint: look at problem 8.6).

8"

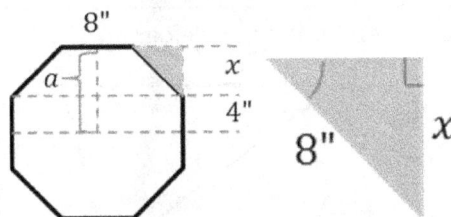

The perimeter is simply 8" × 8 = 64". The apothem drawn from center is $x + 4$ (the 4" because center bisects a side). We draw a right triangle whose upper left angle is $\frac{360°}{8} = 45°$. With a 45-45-90:
$$\frac{x}{1} = \frac{8}{\sqrt{2}} \rightarrow x = 8/\sqrt{2};$$

The area is then:
$$\frac{ap}{2} = \frac{(4+8/\sqrt{2}) \times 64}{2} = 309 \text{ in}^2.$$

8.9: $ABCD$ is a parallelogram

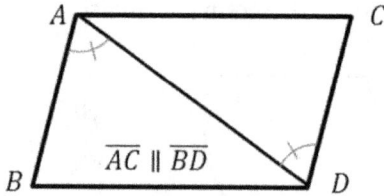

$AC \parallel BD$

Show opposite sides parallel.

Given: $\angle BAD \cong \angle CDA$; $\overline{AC} \parallel \overline{BD}$
Prove: $ABCD$ is a parallelogram

$\angle BAD \cong \angle CDA$	Given
$\overline{AB} \parallel \overline{CD}$	Alt Int $\angle$'s of $\parallel$ lines
$\overline{AC} \parallel \overline{BD}$	Given

$ABCD$ is a parallelogram as opposing sides are parallel.

8.10: $ABCD$ is a parallelogram

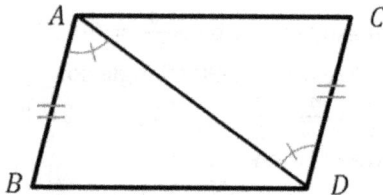

Show a pair of opposite sides are equal and parallel.

Given: $\angle BAD \cong \angle CDA$; $\overline{AB} \cong \overline{CD}$
Prove: $ABCD$ is a parallelogram

$\angle BAD \cong \angle CDA$	Given
$\overline{AB} \parallel \overline{CD}$	Alt Int $\angle$'s of $\parallel$ lines
$\overline{AB} \cong \overline{CD}$	Given

$ABCD$ is a parallelogram as a pair of opposing sides are equal and parallel.

8.11: $ABCD$ is a parallelogram

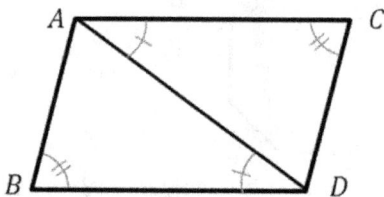

Show opposite sides are equal.

Given: $\angle CAD \cong \angle BDA$; $\angle B \cong \angle C$
Prove: $ABCD$ is a parallelogram

$\angle CAD \cong \angle BDA$	Given
$\angle B \cong \angle C$	Given
$\overline{AD} \cong \overline{AD}$	Reflexive
$\triangle ABD \cong \triangle DCA$	AAS
$\overline{AB} \cong \overline{CD}$	$\triangle$ Congruency
$\overline{AC} \cong \overline{BD}$	$\triangle$ Congruency

$ABCD$ is a parallelogram as opposing sides are equal.

8.12: *ABCD* is a parallelogram

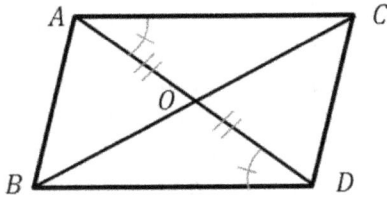

Show a pair of opposite sides are equal and parallel.

Given: $\angle CAO \cong \angle BDO$; $\overline{AO} \cong \overline{OD}$
Prove: *ABCD* is a parallelogram

$\angle CAO \cong \angle BDO$	Given
$\overline{AO} \cong \overline{OD}$	Given
$\angle AOC \cong \angle BOD$	Vertical Angles
$\triangle ABD \cong \triangle DCA$	ASA
$\overline{AC} \cong \overline{BD}$	$\triangle$ Congruency
$\overline{AC} \parallel \overline{BD}$	Alt Int $\angle$'s of $\parallel$ lines

ABCD is a parallelogram as a pair of opposing sides are equal and parallel.

8.13: *ABCD* is a rectangle

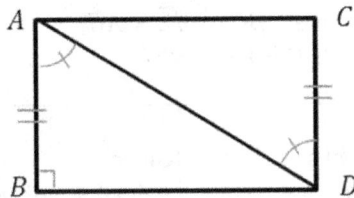

First prove it's a parallelogram, then meet the criteria for a rectangle.

Given: $\angle BAD \cong \angle CDA$; $\overline{AB} \cong \overline{CD}$
$\angle B = 90°$
Prove: *ABCD* is a rectangle

$\angle BAD \cong \angle CDA$	Given
$\overline{AB} \parallel \overline{CD}$	Alt Int $\angle$'s of $\parallel$ lines
$\overline{AB} \cong \overline{CD}$	Given

ABCD is a parallelogram as a pair of opposing sides are equal and parallel.

Since it is given that $\angle B = 90°$, then with at least one right angle, the parallelogram is also a rectangle.

8.14: *ABCD* is a rhombus

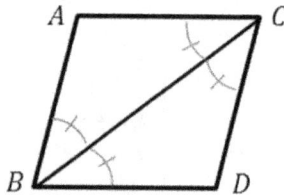

Show it's a parallelogram, then meet the criteria for a rhombus.

Given: $\angle ABC \cong \angle BCD \cong \angle ACB \cong \angle DBC$
Prove: *ABCD* is a rhombus

$\angle ABC \cong \angle BCD$	Given
$\angle ACB \cong \angle DBC$	Given
$\overline{BC} \cong \overline{BC}$	Reflexive
$\triangle ABC \cong \triangle CDB$	ASA
$\overline{AB} \cong \overline{CD}; \overline{AC} \cong \overline{BD}$	$\triangle$ Congruency
ABCD is a Parallelogram	Opp sides equal
$\overline{AB} \cong \overline{CD}$	Opp sides of $\cong \angle$'s in $\triangle ABC$
ABCD is a Rhombus	Pair of $\cong$ adj sides

8.15: *ABCD* is a square

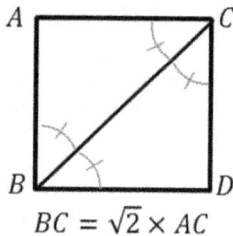

$BC = \sqrt{2} \times AC$

Prove that it's a rhombus AND at least one angle is 90°.

Given: *ABCD* is a rhombus (Use proof from above); $BC = \sqrt{2} \times AC$
Prove: *ABCD* is a square

ABCD is a Rhombus	Given
$\overline{AB} \cong \overline{CD}$	Pair of $\cong$ adj sides
$BC = \sqrt{2} \times AC$	Given
$AB:AC:BC = 1:1:\sqrt{2}$	Summarize above
$\angle A = 90°$	45-45-90 $\triangle$
ABCD is a Square	One angle is 90°

8.16: Using the area formula of a parallelogram is $ap/2$, prove the area of a square is s^2.

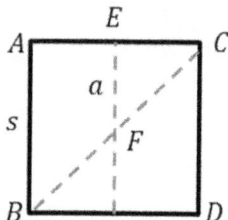

Given: *ABCD* is a square
$AB = BD = DC = AC = s$
Prove: $Area = s^2$

$p = 4s$	Square perimeter
$a = FE = s/2$	As drawn
$Area = ap/2$	Area parallelogram
$Area = \frac{4s(s)}{2 \times 2} = s^2$	Algebra

Chapter 9

Refer to the following figure for Solutions 9.1-9.6:

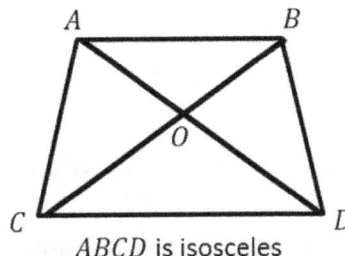

ABCD is isosceles

9.1: Show $\triangle ABC \cong \triangle BAD$	**9.2:** Show $\triangle AOC \cong \triangle BOD$
Since $ABCD$ is an isosceles trapezoid, we know legs and diagonals are congruent: $$AC = BD; \ CB = AD$$ Along with the reflexive statement $AB = AB$, $\triangle ABC \cong \triangle BAD$ by SSS.	Since $\triangle ABC \cong \triangle BAD$, $\angle ACO \cong \angle BDO$. By vertical angles, $\angle AOC \cong \angle BOD$. And, by nature of the isosceles trapezoid, $AC = BD$, $\triangle AOC \cong \triangle BOD$ by AAS.
9.3: Show $AO = OB$ Since $\triangle AOC \cong \triangle BOD$, then by triangle congruency $AO = OB$.	**9.4:** Show $\triangle ACD \cong \triangle BDC$ Since $ABCD$ is an isosceles trapezoid: $$AC = BD; \ AD = CB$$ Along with the reflexive statement $CD = CD$, $\triangle ACD \cong \triangle BDC$ by SSS.
9.5: Show $\triangle COD$ is isosceles Since $\triangle AOC \cong \triangle BOD$, then by triangle congruency $CO = OD$, therefore $\triangle COD$ is isosceles.	**9.6:** Show $\triangle BOA \sim \triangle COD$ By vertical angles, $\angle AOB \cong \angle COD$. We've also shown in Problems 9.3 and 9.5 that $AO = OB$ and $CO = OD$ meaning: $$\frac{AO}{CO} = \frac{OB}{OD}$$ In other words, the sides are in the same proportion making $\triangle BOA \sim \triangle COD$ by SAS.

9.7: Show $ABCE$ is a parallelogram

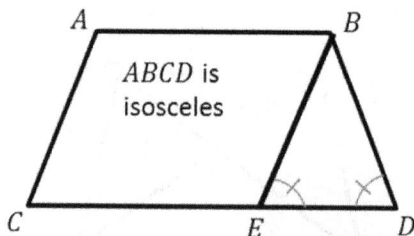

ABCD is isosceles

Because $ABCD$ is an isosceles trapezoid, we know that $\overline{AB} \parallel \overline{CE}$ and $\angle D \cong \angle C$.

From the figure, we see that $\angle D \cong \angle E$ making $\angle C \cong \angle E$ which in turn means $\overline{AC} \parallel \overline{BE}$ (by corresponding angles of parallel lines cut by a transversal).

With two pairs of opposing, parallel sides, $ABCE$ is a parallelogram.

9.8: Show $GF = EH$

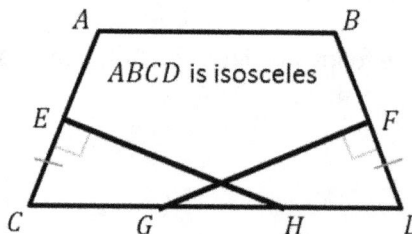

ABCD is isosceles

Because $ABCD$ is an isosceles trapezoid, we know that $\angle D \cong \angle C$.

From the figure, we see that $\angle CEH \cong \angle DFG$ and $EC = FD$.

By ASA, $\triangle CEH \cong \triangle DFG$ which means $GF = EH$.

9.9: Find OP

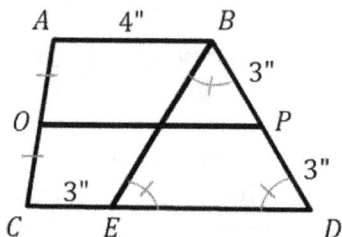

$\triangle BDE$ is equilateral so $ED = 6$" making $CD = 3 + 6 = 9$". By the figure, we see that $AO = OC$ and $BP = PD$, therefore $\overline{OP}$ is a median of trapezoid $ABCD$.

$$OP = \frac{4+9}{2} = 6.5".$$

9.10: Find OP

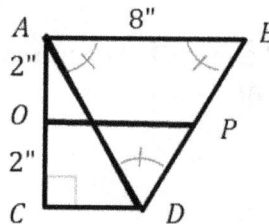

$\overline{OP}$ is a median
of trapezoid $ABCD$

$\triangle ABD$ is equilateral so $AD = 8$" . From the figure, $AC = 4$". $\triangle ACD$ is a right triangle , so by Pythagorean's theorem:
$$CD = \sqrt{8^2 - 4^2} = 6.9"$$

Since $\overline{OP}$ is a median of $ABCD$:
$$OP = \frac{8+6.9}{2} = 7.5".$$

9.11: Find OP

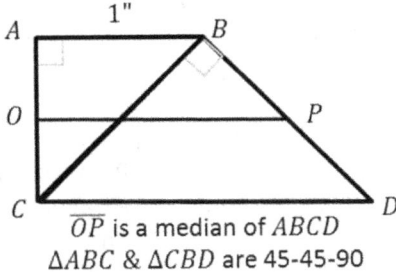

$\overline{OP}$ is a median of $ABCD$
$\triangle ABC$ & $\triangle CBD$ are 45-45-90

Since $\triangle ABD$ is a 45-45-90 triangle with side ratio $1:1:\sqrt{2}$, $BC = \sqrt{2}"$.

Since $\triangle CBD$ is also a 45-45-90 triangle, $CD = \sqrt{2} \times \sqrt{2} = 4"$.

Since $\overline{OP}$ is a median of $ABCD$:

$$OP = \frac{4+1}{2} = 2.5".$$

9.12: Find BD

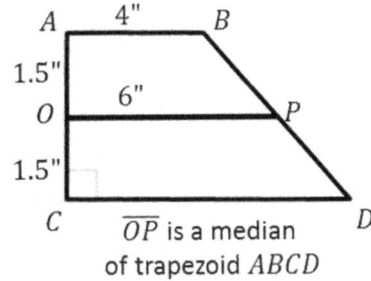

$\overline{OP}$ is a median of trapezoid $ABCD$

Since $\overline{OP}$ is a median of $ABCD$:

$$6 = \frac{4+CD}{2} \rightarrow CD = 8".$$

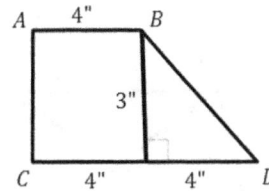

If we draw a perpendicular from vertex B to the base, we have a rectangle with bottom side 4" and a right triangle with sides 3" and 4". By Pythagorean's (or 3:4:5 triplet), we have $BD = 5"$.

9.13:

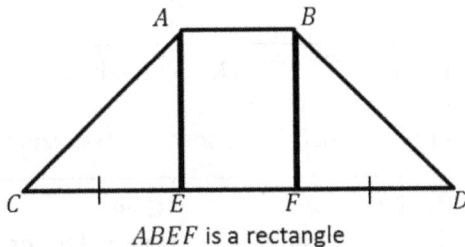

$ABEF$ is a rectangle

Prove bases are parallel (trapezoid property), then $\triangle ACE \cong \triangle BDF$ which proves $AC = BD$ (isosceles property).

Given: $ABEF$ is a rectangle; $CE = FD$
Prove: $ABCD$ is an isosceles trapezoid

$ABEF$ is a Rectangle	Given
$\overline{AE} \cong \overline{BF}$	Rect. property
$\angle AEF \cong \angle BFE \cong 90°$	Rect. property
$\angle AEC \cong \angle BFD \cong 90°$	Supplement $\angle$'s
$\overline{AB} \parallel \overline{EF} \parallel \overline{CD}$	Rect. property
$ABCD$ is a trapezoid	Parallel bases
$\overline{CE} \cong \overline{FD}$	Given
$\triangle ACE \cong \triangle BDF$	SAS
$AC = BD$	$\triangle$ Congruency
$ABCD$ is Isosc. Trap.	Equal legs

9.14:

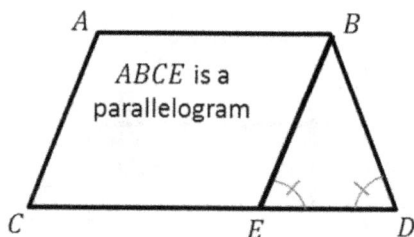

ABCE is a parallelogram

Prove parallel bases, then equal legs.

Given: $ABCE$ is a parallelogram;
$\angle BED \cong \angle BDE$

Prove: $ABCD$ is an isosceles trapezoid

$ABCE$ is a Parallelogram	Given
$\overline{AB} \parallel \overline{CE} \parallel \overline{CD}$	Parall. property
$ABCD$ is a trapezoid	Parallel bases
$\overline{AC} \cong \overline{BE}$	Parall. property
$\angle BED \cong \angle BDE$	Given
$\overline{BD} \cong \overline{BE}$	Isosc. Triangle
$\overline{AC} \cong \overline{BD}$	Transitive
$ABCD$ is Isosc. Trap.	Equal legs

9.15:

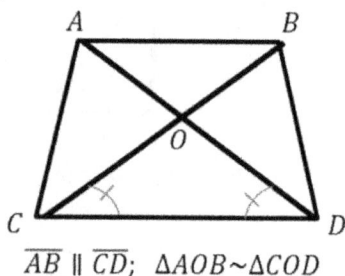

$\overline{AB} \parallel \overline{CD}$; $\triangle AOB \sim \triangle COD$

Prove diagonals are equal.

Given: $\overline{AB} \parallel \overline{CD}$; $\triangle AOB \sim \triangle COD$;
$\triangle COD$ is isosceles

Prove: $ABCD$ is an isosceles trapezoid

$\overline{AB} \parallel \overline{CD}$	Given
$ABCD$ is a trapezoid	Parallel bases
$\triangle AOB \sim \triangle COD$	Given
$\triangle COD$ is isosceles	Given
$CO = OD$	Isosc. Triangle
$AO = OB$	Similar Triangles
$CO + OB = AO + OD$	Add lengths
$ABCD$ is Isosc. Trap.	Equal diagonals

9.16:

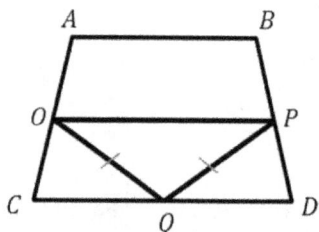

$\overline{AB} \parallel \overline{CD}$; O, P, Q are bisection points

Prove equal base angles.

Given: $\overline{AB} \parallel \overline{CD}$; O, P, Q are bis. pts.;
$OQ = QP$

Prove: $ABCD$ is an isosceles trapezoid

$\overline{AB} \parallel \overline{CD}$	Given
$ABCD$ is a trapezoid	Parallel bases
$OC = PD; CQ = QD$	Given
$OQ = QP$	Given
$\triangle OCQ \cong \triangle PDQ$	SSS
$\angle OCQ \cong \angle PDQ$	$\triangle$ Congruency
$ABCD$ is Isosc. Trap.	Equal base $\angle$'s

Chapter 10

10.1: Find $m\overline{OQ}$:

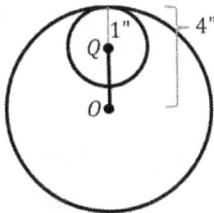

$\odot Q$ is tangent to $\odot O$.

The distance from point Q to the top of its circle is 1"; The distance from point O to the top of its circle is 4". Because $\odot Q$ is tangent to $\odot O$, the top of both circles are the same point. Therefore,

$$\overline{OQ} = 4 - 1 = 3".$$

10.2: Find $m\overline{AB}$:

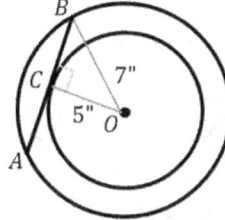

$\overline{AB}$ is tangent to inner circle; Circles are concentric.

Redrawing the radii as shown above, and knowing that a radius to a tangent is a perpendicular bisector, we can use Pythagorean's rule for a right triangle:

$$BC = \sqrt{7^2 - 5^2} = 4.9";$$
$$AC = 2 \times BC = 9.8".$$

10.3: Find the area of the pentagram.

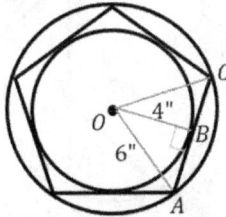

Circles are concentric; Pentagon is regular, inscribed, and circumscribed.

Redrawing the radii and recognizing $\overline{AC}$ is a tangent chord, we see that the apothem is 4". For ΔABO:
$$AC = 2AB = 2\sqrt{6^2 - 4^2} = 8.9";$$

The perimeter of the pentagram is then $5 \times 8.9 = 44.7"$ making the area:

$$\frac{ap}{2} = \frac{4(44.7)}{2} = 89.4 \text{ in}^2.$$

10.4: Find the area of the hexagram.

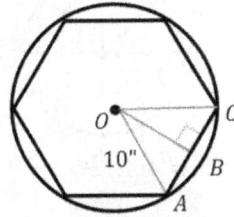

Regular hexagon is inscribed in circle of diameter 20".

The interior angles of a hexagram measure $\frac{360}{6} = 60°$. Drawing $\overline{OB}$, we get $\angle AOB = 30°$ and ΔOAB is a 30-60-90:

The apothem is $\frac{OB}{\sqrt{3}} = \frac{10}{2} \rightarrow OB = 8.7"$;

$$AC = 2AB; \frac{AB}{1} = \frac{10}{2} \rightarrow AC = 10"$$

$$\frac{ap}{2} = \frac{(8.7)(6)(10)}{2} = 260 \text{ in}^2.$$

10.5:

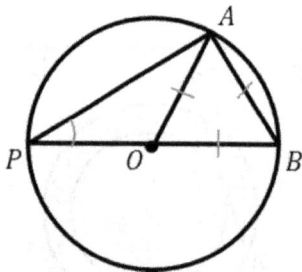

Because $\triangle AOB$ is equilateral,

$$\angle AOB = \overset{\frown}{AOB} = 60°;$$

$$\angle P = \frac{\overset{\frown}{AOB}}{2} = 30°.$$

10.6:

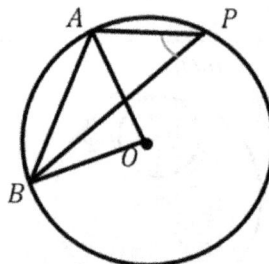

$$m\angle BAO = 40°.$$

Because $\triangle AOB$ is isosceles,

$$\angle BAO = \angle ABO = 40°;$$
$$\overset{\frown}{AOB} = 180 - 40 - 40 = 100°;$$

$$\angle P = \frac{\overset{\frown}{AOB}}{2} = 50°.$$

10.7:

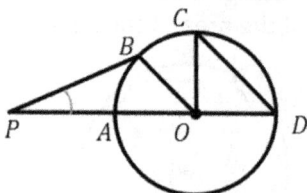

$$m\angle OCD = 45°; \quad \overline{BO} \parallel \overline{CD}.$$

Because $\triangle COD$ is isosceles,

$$\angle ODC = \angle OCD = 45°;$$
$$\overset{\frown}{COD} = 180 - 45 - 45 = 90°;$$

Since $\overline{BO} \parallel \overline{CD}$, $\angle AOB = \angle ODC = 45°$ by corresponding angles.

$$\angle P = \frac{\overset{\frown}{COD} - \overset{\frown}{AOB}}{2} = \frac{90 - 45}{2} = 22.5°.$$

10.8:

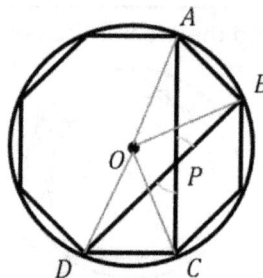

Regular octagon is inscribed.

The interior angles of an octagon are

$$\frac{360}{8} = 45°, \text{ so } \overset{\frown}{AOB} = \overset{\frown}{COD} = 45°;$$

$$\angle P = \frac{\overset{\frown}{COD} + \overset{\frown}{AOB}}{2} = \frac{45 - 45}{2} = 45°.$$

10.9:

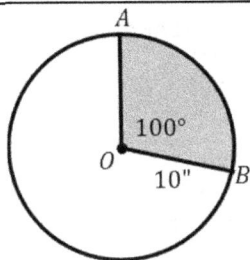

The area of the whole circle is:
$$\pi r^2 = \pi 10^2 = 314 \text{ in}^2;$$

The area of sector AOB is:
$$\frac{100}{360}(314) = 87.3 \text{ in}^2.$$

10.10:

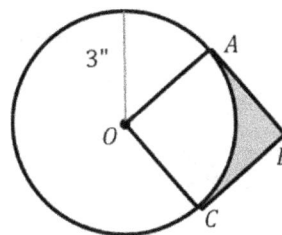

Shape $OABC$ is a square.

The area of the shaded region is the area of the square minus that of the sector.

The area of the square is $3^2 = 9 \text{ in}^2$;
The area of the sector is:
$$\frac{90}{360}\pi r^2 = \frac{90}{360}\pi 3^2 = 7.1 \text{ in}^2;$$

The area of the shaded region is:
$$9 - 7.1 = 1.9 \text{ in}^2.$$

10.11:

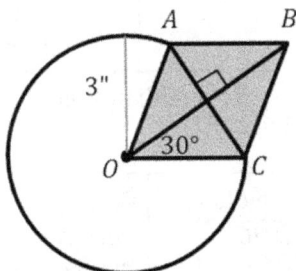

Shape $OABC$ is a rhombus.

Drawing chord $\overline{AC}$ and its perpendicular bisector $\overline{OB}$, we get four 30-60-90 triangles: For the lower triangle:

$$\frac{OB/2}{\sqrt{3}} = \frac{OC}{2} = \frac{3}{2} \rightarrow \frac{OB}{2} = 2.6" = a;$$

The perimeter is $4(3) = 12"$;
Area of $OABC = \frac{ap}{2} = \frac{2.6(12)}{2} = 15.6 \text{ in}^2.$

10.12:

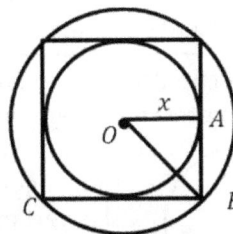

Circles are concentric; Square is Inscribed and circumscribed.

The area of the inner circle is πx^2. Side $\overline{CB}$ measures $2x$, so the area of the square is $4x^2$.

To find the radius of the larger circle:
$$OA = AB = x; \quad OB = \sqrt{x^2 + x^2} = \sqrt{2}x;$$

The area of the large circle is then $2\pi x^2$ or twice that of the smaller circle.

10.13: $\overline{OP} \cong \overline{PQ}$

$\overline{AB}$ is tangent to $\odot O$ and $\odot Q$

Recognize radius is perpendicular to the tangent, then prove congruent triangles.

Given: $\overline{AB}$ is tangent to $\odot O$ and $\odot Q$; $\overline{OA} \cong \overline{BQ}$
Prove: $\overline{OP} \cong \overline{PQ}$

$\overline{AB}$ is tangent to $\odot O$ and $\odot Q$	Given
$\overline{OA} \perp \overline{AB}$; $\overline{QB} \perp \overline{AB}$	Radius to Tang.
$\overline{OA} \cong \overline{BQ}$	Given
$\angle OPA \cong \angle BPQ$	Vertical Angles
$\triangle OPA \cong \triangle QPB$	AAS
$\overline{OP} \cong \overline{PQ}$	$\triangle$ Congruency

10.14: $\overline{OE} \cong \overline{OF}$

$\overline{AB} \parallel \overline{CD}$.

Use the property of ∥chords w/ the equivalent statements of $\cong$ arcs & chords.

Given: $\overline{AB} \parallel \overline{CD}$
Prove: $\overline{OE} \cong \overline{OF}$

$\overline{AB} \parallel \overline{CD}$	Given
$\overset{\frown}{AOD} \cong \overset{\frown}{BOC}$	$\cong$ arcs between ∥chords
$\overline{AD} \cong \overline{BC}$	$\cong$ arcs $\to$ $\cong$ chords
$\overline{OE} \cong \overline{OF}$	$\cong$ chords equidistant from center

10.15: The chords between two diameters are congruent.

Draw figure;
Use vertical angles to get equal arcs.

Given: $\overline{AC}, \overline{BD}$ are diameters
Prove: $\overline{AB} \cong \overline{DC}$

$\overline{AC}, \overline{BD}$ are diam.	Given
$\overset{\frown}{AOB} \cong \overset{\frown}{DOC}$	Vert. Angles
$\overline{AB} \cong \overline{DC}$	$\cong$ arcs $\to$ $\cong$ chords

10.16: The only perpendicular from center circle to a tangent line is at the point of intersection between the tangent and the circle.

Draw figure;
Use proof by contradiction.

Given: $\overline{OA} \perp \overline{AC}$
Prove: $\overline{OB} \not\perp \overline{AC}$

$\overline{OA} \perp \overline{AC}$	Given
$\overline{OB} \perp \overline{AC}$	Assume
$\angle O = 180 - 90 - 90$	$\triangle AOB$ Angle Sum
$\angle O = 0°$	Arithmetic

Either point B is point A, or $\overline{OB} \not\perp \overline{AC}$

Chapter 11

11.1: Point symmetry.	11.2: Line symmetry.
11.3: Multiple line & rotation symmetries.	11.4: Line, point, & rotation symmetries.
11.5: One translation Or, two reflections.	11.6: One reflection.

11.7:

One rotation

Or two reflections.

11.8:

One translation

Or two reflections.

Perform the transformations indicated on the following triangle for Problems 11.9-11.16:

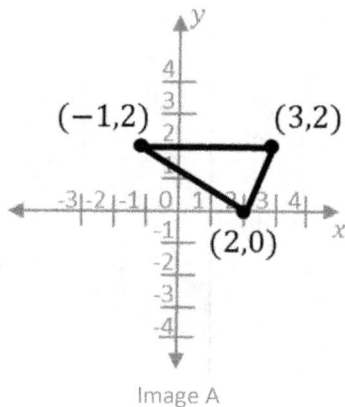

Image A

11.9: $T_{-2,3}(A)$

For each point, we translate as follows:
$$(A'_x, A'_y) = ([A_x + h], [A_y + k])$$

$$(-1,2) \rightarrow ([-1-2], [2+3]) = (-3,5)$$
$$(3,2) \rightarrow ([3-2], [2+3]) = (1,5)$$
$$(2,0) \rightarrow ([2-2], [0+3]) = (0,3)$$

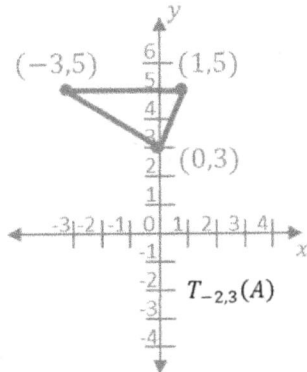

11.10: $r_{x=1}(A)$

For each point, we reflect as follows:
$$(A'_x, A'_y) = ([2a - A_x], A_y)$$

$$(-1,2) \rightarrow ([2-(-1)], 2) = (3,2)$$
$$(3,2) \rightarrow ([2-3], 2) = (-1,2)$$
$$(2,0) \rightarrow ([2-2], 0) = (0,0)$$

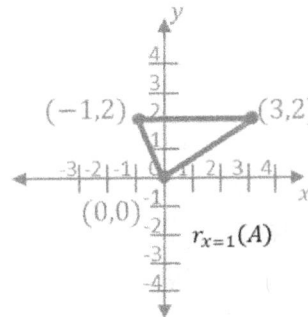

11.11: $r_{y=-1}(A)$

For each point, we reflect as follows:
$$(A'_x, A'_y) = (A_x, [2b - A_y])$$

$$(-1,2) \rightarrow (-1, [-2-2]) = (-1,-4)$$
$$(3,2) \rightarrow (3, [-2-2]) = (3,-4)$$
$$(2,0) \rightarrow (2, [-2-0]) = (2,-2)$$

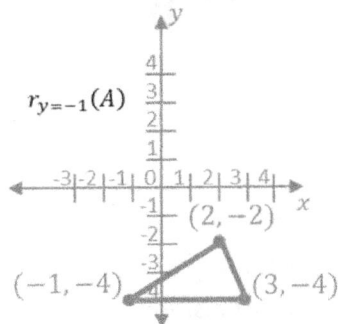

11.12: $r_{(1,1)}(A)$

For each point, we reflect as follows:
$$(A'_x, A'_y) = ([2a - A_x], [2b - A_y])$$

$$(-1,2) \rightarrow ([2+1], [2-2]) = (3,0)$$
$$(3,2) \rightarrow ([2-3], [2-2]) = (-1,0)$$
$$(2,0) \rightarrow ([2-2], [2-0]) = (0,2)$$

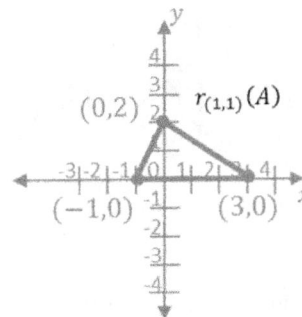

11.13: $D_2(A)$

For each point, we dilate as follows:
$$\left(A'_x, A'_y\right) = \left([cA_x], [cA_y]\right)$$

$(-1,2) \rightarrow (-2,4);\ (3,2) \rightarrow (6,4);$
$(2,0) \rightarrow (4,0)$

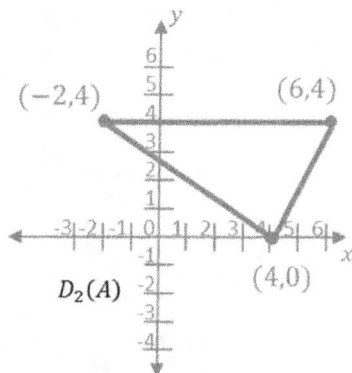

11.14: $R_{90°}(A)$

For each point, we rotate CCW as follows:
$$\left(A'_x, A'_y\right) = \left(-A_y, A_x\right)$$

$(-1,2) \rightarrow (-2,-1);\ (3,2) \rightarrow (-2,3);$
$(2,0) \rightarrow (0,2)$

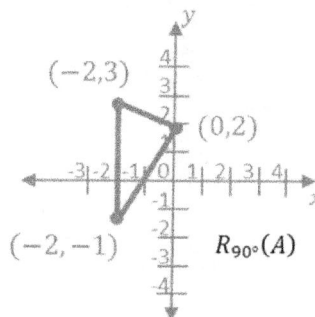

11.15: $T_{1,0} \circ R_{90°}(A)$

Rotate, then translate:
$(-1,2) \rightarrow (-2,-1) \rightarrow (-1,-1)$
$(3,2) \rightarrow (-2,3) \rightarrow (-1,3)$
$(2,0) \rightarrow (0,2) \rightarrow (1,2)$

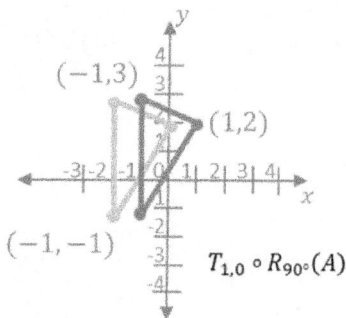

First transformation is
shown in light grey.

11.16: $R_{90°} \circ T_{1,0}(A)$

Translate, then rotate:
$(-1,2) \rightarrow (0,2) \rightarrow (-2,0)$
$(3,2) \rightarrow (4,2) \rightarrow (-2,4)$
$(2,0) \rightarrow (3,0) \rightarrow (0,3)$

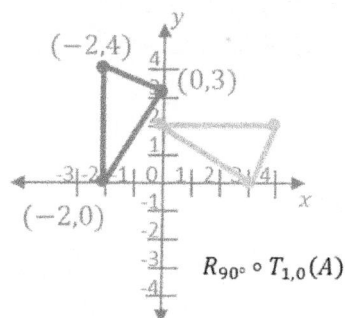

First transformation is
shown in light grey.

Chapter 12

12.1: 1" ⊥ distance from the line and point as shown below.

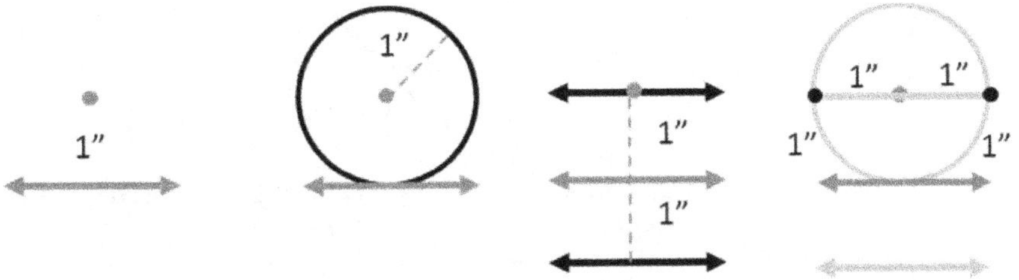

1"	1"	1" 1" 1"	1" 1" 1" 1"
Original	Locus about point	Locus about line	There are 2 loci

12.2: 1" ⊥ distance from two lines and a point as shown below.

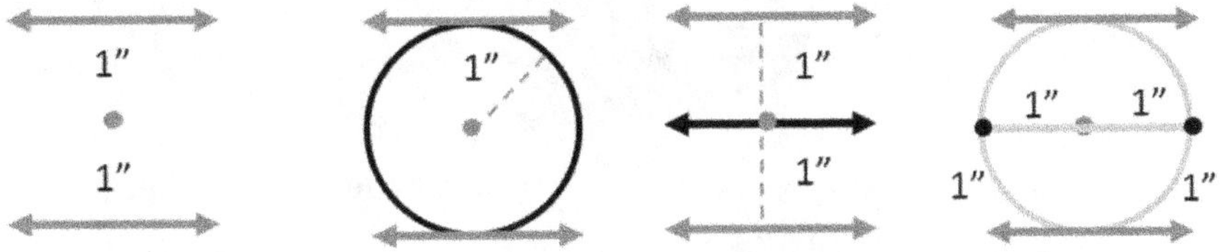

1" 1"	1"	1" 1"	1" 1" 1" 1"
Original	Locus about point	Locus about lines	There are 2 loci

12.3: Equidistant from the two points and 1" from the line

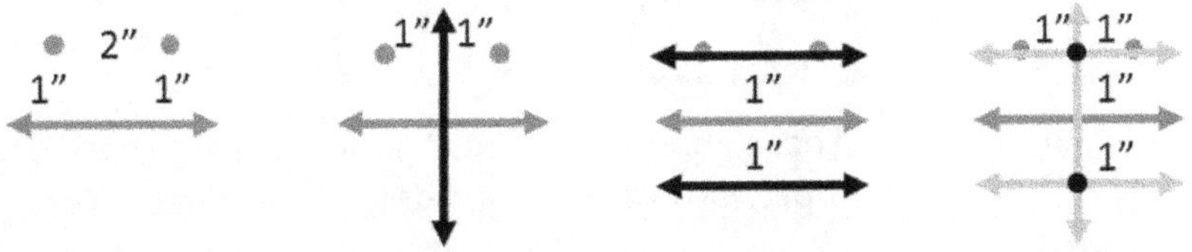

2" 1" 1"	1" 1"	1" 1"	1" 1" 1" 1"
Original	Locus about 2 points	Locus about line	There are 2 loci

12.4: Equidistant from four pints outlining a 2" square

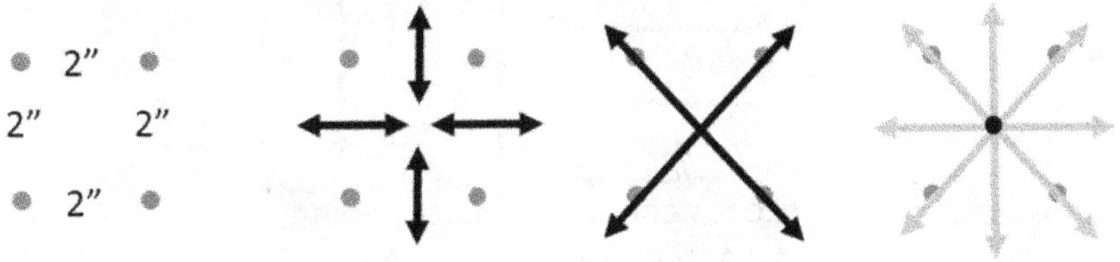

2" 2" 2" 2"			
Original	Locus about adjacent points	Locus about diagonal points	There is 1 loci

12.5: Equidistant from the two lines and 1" from the point

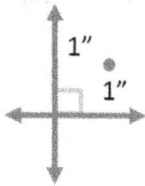

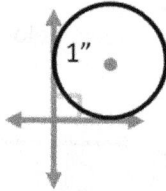

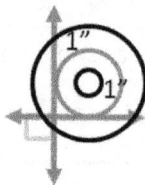

| Original | Locus about point | Locus about lines | There are 2 loci |

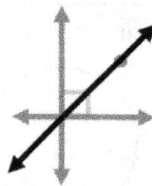

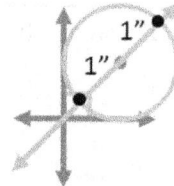

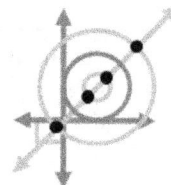

12.6: Equidistant from the two lines and 1" from the circle

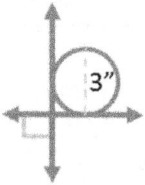

| Original | Locus about point | Locus about lines | There are 4 loci |

12.7: 1" ⊥ distance from the sides of a 2" square

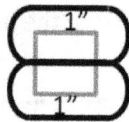

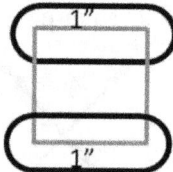

| Original | Locus about top& bottom | Locus about sides | There is 1 loci |

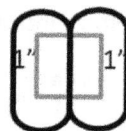

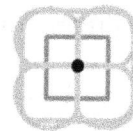

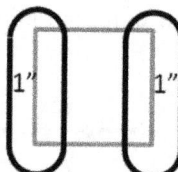

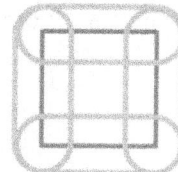

12.8: 1" ⊥ distance from the sides of a 4" square

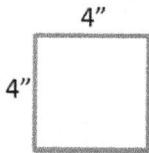

| Original | Locus about top& bottom | Locus about sides | There are no loci |

12.9: Centroid

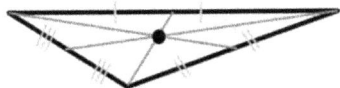

The intersection of medians.

12.10: Orthocenter

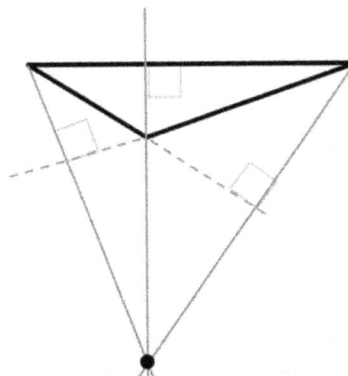

The intersection of altitudes.

12.11: Circumcenter

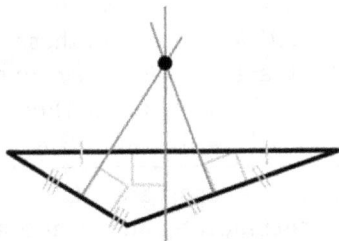

The intersection of perpendicular bisectors.

12.12: Incenter

The intersection of angle bisectors.

12.13: How far from the top vertex is the centroid of the equilateral triangle given below?

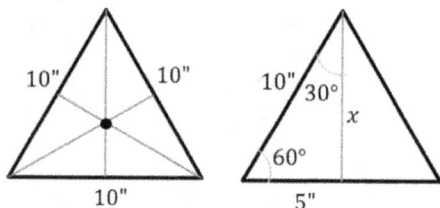

We know the centroid is 2/3 the median length from the vertex. Considering the vertical median, we have a 30-60-90 triangle, which gives us:

$$\frac{x}{\sqrt{3}} = \frac{10}{2} \rightarrow x = 5\sqrt{3} = 8.7"$$

The centroid is then located:

$$\frac{2x}{3} = 5.8" \text{ from the vertex.}$$

12.14: Is it true that point O is the orthocenter of $\triangle MNO$ and the circumcenter of $\triangle ABC$ if both triangles are 45-45-90.?

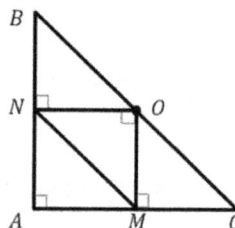

We know that the orthocenter is at the vertex of a right triangle, so point O is the orthocenter of $\triangle MNO$. The circumcenter of $\triangle ABC$ must lie on the hypotenuse. We can see that $\angle BNO$ and $\angle CMO$ are both 90° (which can be confirmed with a series of $\triangle$ angle sums). What remains is to confirm that $BN = NA, AM = AC$, and $BO = OC$. By AAA, we have $\triangle ANM \sim \triangle NBO \sim \triangle MOC \sim \triangle ONM$. Since all of these triangles share like sides, they are not only similar, but congruent. Therefore, $BN = NA, AM = AC$, and $BO = OC$.

12.15: For an equilateral triangle, which centers share the same location?

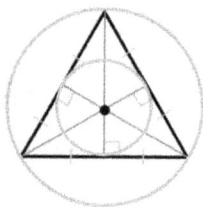

Because the medians of an equilateral triangle are also angle bisectors, perpendicular side bisectors, and altitudes, all of the centers are the same.

12.16: How far apart are the circumcenter and orthocenter of a right isosceles triangle with an 8" hypotenuse?

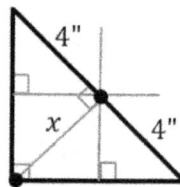

The question asks the length of the median. Looking at the bottom 45-45-90 triangle with sides x and 4", one can see that $x = 4"$.

Chapter 13

13.1: Find S and V:

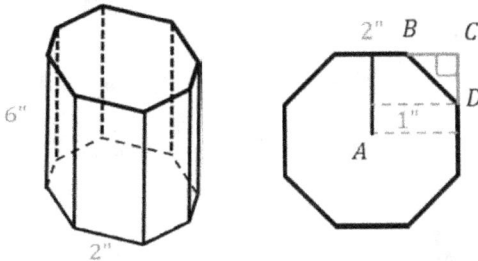

The perimeter is $p = 8(2) = 16$", and the apothem, a, is:

$$1 + CD = 1 + \frac{2}{\sqrt{2}} = 2.4" \text{ (see Prob. 8.8)};$$

Therefore, $B = \frac{ap}{2} = \frac{16(2.4)}{2} = 19.3 \text{ in}^2;$

$$S = 2B + ph = 135 \text{ in}^2;$$

$$V = Bh = 116 \text{ in}^3.$$

13.2: Find S and V:

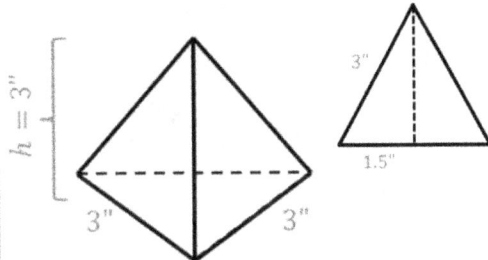

The perimeter is $p = 3(3) = 9$". The triangle base, b, is 3", and the triangle height, H, is $\sqrt{3^2 - 1.5^2} = 2.6$";

Therefore, $B = \frac{bH}{2} = \frac{3(2.6)}{2} = 3.9 \text{ in}^2;$

$$S = B + \frac{ph}{2} = 3.9 + \frac{9(3)}{2} = 17.4 \text{ in}^2;$$

$$V = \frac{Bh}{3} = 3.9 \text{ in}^3.$$

13.3: Compare the volumes of the two solids below to the solid of Problem 13.1 (V_0).

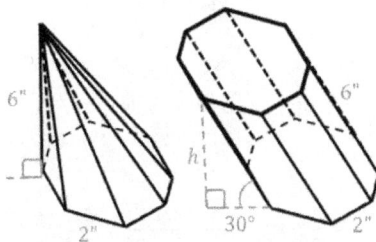

B is the same, so for the first figure:

$$V_1 = \frac{Bh}{3} = \frac{19.3(6)}{3} = 38.6 \text{ in}^3, \text{ or } V_0/3.$$

For the second, $\frac{h}{1} = \frac{6}{2} \rightarrow h = 3$";

$$V_2 = Bh = 19.3(3) = 57.9 \text{ in}^3, \text{ or } V_0/2.$$

13.4: Compare the volumes of the two solids below to the solid of Problem 13.2 (V_0).

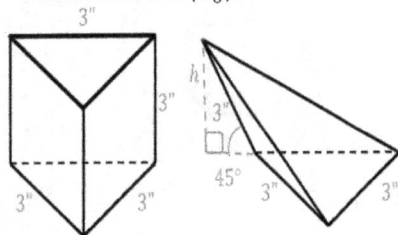

B is the same, so for the first figure:

$$V_1 = Bh = 3.9(3) = 11.7 \text{ in}^3, \text{ or } 3V_0.$$

For the second, $\frac{h}{1} = \frac{3}{\sqrt{2}} \rightarrow h = 2.1$";

$$V_2 = \frac{Bh}{3} = \frac{3.9(2.1)}{3} = 2.8 \text{ in}^3, \text{ or } V_0/\sqrt{2}.$$

13.5:

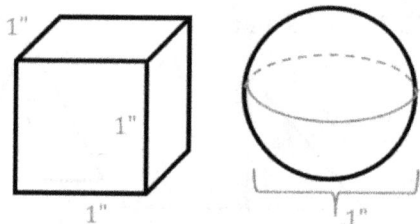

$$V_1 = Bh = 1(1)(1) = 1 \text{ in}^3;$$

$$V_2 = \frac{4}{3}\pi r^3 = \frac{4\pi(0.5)^3}{3} = 0.5 \text{ in}^3, \text{ or } \sim \frac{V_1}{2}.$$

13.6:

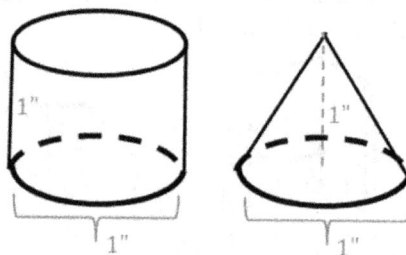

$$V_1 = Bh = \pi(0.5)^2(1) = 0.78 \text{ in}^3;$$

$$V_2 = \frac{Bh}{3} = \frac{\pi(0.5)^2(1)}{3} = 0.26 \text{ in}^3, \text{ or } \frac{V_1}{3}.$$

13.7: A cube with side x and a cube with side $2x$.

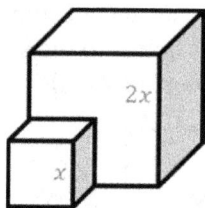

$$V_1 = Bh = x(x)(x) = x^3 \text{ in}^3;$$

$$V_2 = Bh = 2x(2x)(2x) = 8x^3 \text{ in}^3,$$

or $8V_1$.

13.8: A sphere with diameter x and a sphere with diamter $2x$.

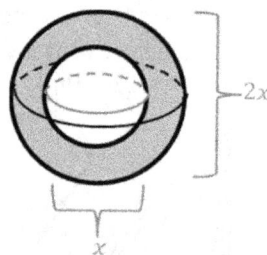

$$V_1 = \frac{4}{3}\pi r^3 = \frac{4\pi(x/2)^3}{3} = \frac{4\pi}{24}x^3 \text{ in}^3;$$

$$V_2 = \frac{4}{3}\pi r^3 = \frac{4\pi}{3}x^3 \text{ in}^3, \text{ or } 8V_1.$$

13.9: If the height is doubled or the radius is doubled?	**13.10:** If the height is doubled or the radius is doubled?

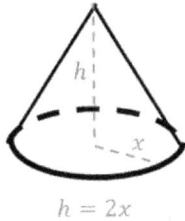

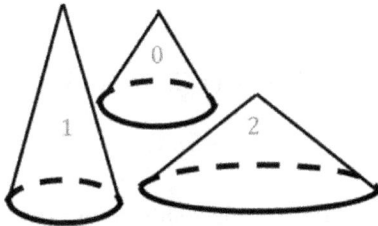

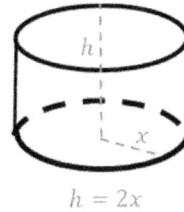

$h = 2x$

$h = 2x$

Case 1: Let $h = 4x$:

$$S_1 = B + \frac{ph}{2} = \pi x^2 + \frac{(2\pi x)4x}{2} = 5\pi x^2;$$

$$V_1 = \frac{Bh}{3} = \frac{\pi x^2(4x)}{3} = \frac{4\pi}{3}x^3.$$

Case 2: Let $r = 2x$:

$$S_2 = \pi(2x)^2 + \frac{(2\pi(2x))2x}{2} = 8\pi x^2;$$

$$V_2 = \frac{\pi(2x)^2(2x)}{3} = \frac{8\pi}{3}x^3.$$

Both surface area and volume are larger for the case of doubling the radius.

Case 1: Let $h = 4x$:

$$S_1 = B + ph = \pi x^2 + (2\pi x)4x$$

$$= 9\pi x^2;$$

$$V_1 = Bh = \pi x^2(4x) = 4\pi x^3.$$

Case 2: Let $r = 2x$:

$$S_2 = \pi(2x)^2 + (2\pi(2x))2x = 12\pi x^2;$$

$$V_2 = \pi(2x)^2(2x) = 8\pi x^3.$$

Both surface area and volume are larger for the case of doubling the radius.

13.11:

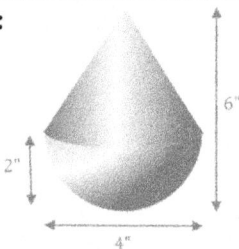

The volume is half a sphere plus a cone:

$$V = \frac{4}{6}\pi r^3 + \frac{Bh}{3}$$

$$= \frac{4\pi 2^3}{6} + \frac{(\pi 2^2)(6-2)}{3} = 33.5 \text{ in}^3.$$

The surface area is half a sphere plus the *lateral face* of the cone:

$$S = 2\pi r^2 + \frac{ph}{2}$$

$$= 2\pi 2^2 + \frac{2\pi(2)(6-2)}{2} = 50.2 \text{ in}^2.$$

13.12:

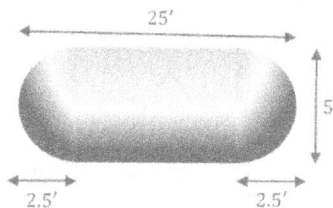

The volume is a sphere plus a cylinder:

$$V = \frac{4}{3}\pi r^3 + Bh$$

$$= \frac{4\pi(2.5)^3}{3} + (\pi(2.5)^2)(25-5)$$

$$= 458 \text{ ft}^3.$$

The surface area is a sphere plus the *lateral face* of the cylinder:

$$S = 4\pi r^2 + ph$$

$$= 4\pi(2.5)^2 + 2\pi(2.5)(25-5)$$

$$= 39.3 \text{ ft}^2.$$

13.13: Will a sphere with a circumference of 10" fit inside a cube with volume 30 in³?

The cube side is:

$$V_{cube} = Bh = s^3 = 30$$

$$s = \sqrt[3]{30} = 3.1"$$

The diameter of the sphere is:

$$p = 2\pi r = 10$$

$$2r = \frac{10}{\pi} = 3.2"$$

No, the diameter of the sphere exceeds the height of the box.

13.14: How long is a string of 10 balls laid end to end each with a volume of 9 in³?

Each sphere has a diameter of:

$$V_{sphere} = \frac{4}{3}\pi r^3 = 9$$

$$r = \sqrt[3]{\frac{9(3)}{4\pi}} = 1.3"$$

$$Diameter = 2r = 2.6"$$

10 such spheres then measures

$$10(2.6) = 26".$$

13.15: Which is taller: A sphere or cube each with a volume of 100 in³?

The cube side is:

$$V_{cube} = Bh = s^3 = 100$$

$$s = \sqrt[3]{100} = 4.6"$$

Each sphere has a diameter of:

$$V_{sphere} = \frac{4}{3}\pi r^3 = 100$$

$$r = \sqrt[3]{\frac{100(3)}{4\pi}} = 2.9"$$

$$Diameter = 2r = 5.8"$$

The sphere is taller

13.16: If a cone has a 10" diameter base and is 10" high, what is the diameter of a 10" high cylinder with the same volume?

The cone volume is:

$$V_{cone} = \frac{Bh}{3} = \frac{\pi 5^2 (10)}{3} = 262 \text{ in}^3$$

The cylinder diameter is then:

$$V_{cyl} = Bh = \pi r^2 (10) = 262$$

$$r = \sqrt{\frac{262}{10\pi}} = 2.9"$$

$$Diameter = 2r = 5.8"$$

Appendix D: Additional Proofs

If two sides of a triangle are not congruent:
- **The opposing angles are not congruent, and**
- **The larger angle is opposite the larger side.**

For the proof, we will draw a line $\overline{BO}$ that intersects $\overline{AC}$ such that $\overline{AB} \cong \overline{AO}$. Therefore, ΔABO is an isosceles triangle.

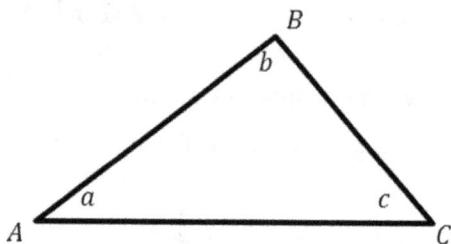

Triangle for Proof.

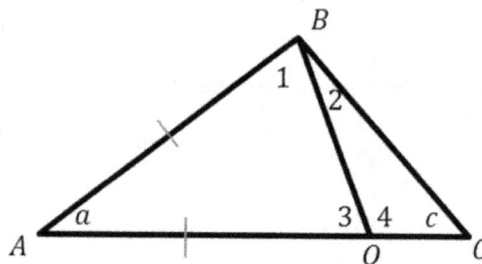

Modified triangle for Proof.

Given: $AC > AB$; $\overline{AB} \cong \overline{AO}$
Prove: $m\angle b > m\angle c$

Statement	Justification
$m\angle 1 = m\angle 3$	Isosceles Δ Equality
$m\angle 3 = m\angle 2 + m\angle c$	Exterior Angles
$m\angle 1 = m\angle 2 + m\angle c$	Substitution
$m\angle c = m\angle 1 - m\angle 2$	Algebra
$m\angle b = m\angle 1 + m\angle 2$	Angle Summation
$m\angle b > m\angle c$	Since $m\angle 2 > 0$ and comparing prior two lines of proof

If two angles of a triangle are not congruent:
- **The opposing sides are not congruent, and**
- **The larger side is opposite the larger angle.**

We will do a proof by contradiction.

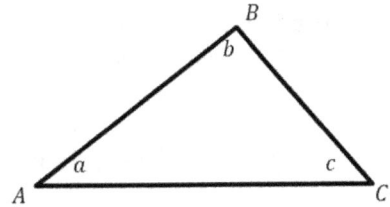

Triangle for Proof.

Given: $m\angle b > m\angle c$; $\overline{AB} \cong \overline{AO}$
Prove: $AC > AB$

Statement	Justification
$AC = AB$	Contradiction assertion
$m\angle b = m\angle c$	Isosceles equality
$m\angle b > m\angle c$	Given
$AC \neq AB$	Contradiction of given
$AC < AB$	Contradiction assertion
$m\angle b < m\angle c$	Theorem on prior page
$m\angle b > m\angle c$	Given
$AC \not< AB$	Contradiction of given
$AC > AB$	Only option left

**The following statements are equivalent
for the same or congruent circles:**
- **Two chords are equidistant from the center**
- **Two chords are congruent**
- **Two arcs defined by such chords are congruent**

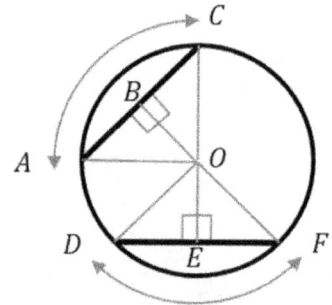

Given: $\odot O$
Prove: $OB = OE \leftrightarrow AC = DF \leftrightarrow \angle AOC \cong \angle DOF$

Two chords in a circle.

Statement	Justification
$OB = OE;\ OA = OC = OD = OF$	Assume Given; Radii
$\triangle AOB \cong \triangle BOC \cong \triangle DOE \cong \triangle EOF$	Exception to SSA for Right $\triangle$'s
$AB = BC = DE = EF$	Triangle Congruence
$AC = AB + BC;\ DF = DE + EF$	Addition of Line Segments
$AC = DF$	Substitution
$AC = DF; OA = OC = OD = OF$	Assume Given; Radii
$\triangle AOC \cong \triangle DOF$	SSS Congruency
$\angle AOC \cong \angle DOF;\ OB = OE$	Triangle Congruency
$\angle AOC \cong \angle DOF;\ OA = OC = OD = OF$	Assume Given; Radii
$\triangle AOC \cong \triangle DOF$	SAS Congruency
$OB = OE;\ AC = DF$	Triangle Congruency

Appendix E: Introduction to Trigonometry

There are some geometry courses that require a familiarity with elementary trigonometry. Trigonometry is the study of the relationships of sides and angles in right triangles. For completion, some basic concepts are outlined below.

Triangles must obey some rigid properties that allow three sides to form a closed figure. Right triangles have the added property that one angle is 90° and the other two angles are complimentary. In the figure to the right, we see an acute angle x within a right triangle. The sides are labeled: the hypotenuse (HYP), the other side next to or adjacent to the angle (ADJ), and the side opposite the angle (OPP).

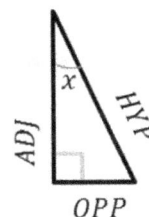

General right triangle.

Imagine keeping the ADJ side the same while increasing the angle x. In order to keep a closed right triangle, the OPP and HYP sides must grow larger, and the third angle must get smaller. Some mathematician long ago discovered that these changes are very predictable. If we take the ratio of any two sides of a right triangle, we can predict the angle x. The ratios are given the special names **sine**, **cosine**, and **tangent**, as follows:

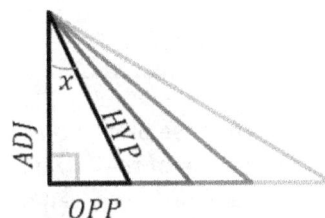

Right triangle with expanding angle x.

Term	Notation	Equation
Sine	$\sin(x)$	$\sin(x) = \dfrac{OPP}{HYP}$
Cosine	$\cos(x)$	$\cos(x) = \dfrac{ADJ}{HYP}$
Tangent	$\tan(x)$	$\tan(x) = \dfrac{\sin(x)}{\cos(x)} = \dfrac{OPP}{ADJ}$

Below are plots of the ratios (the x-axis is in degrees):

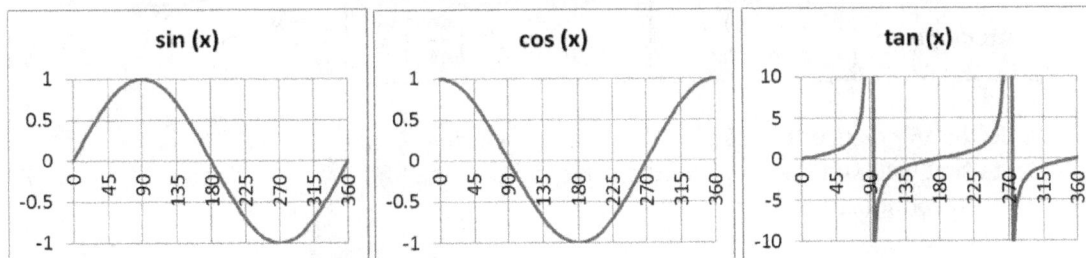

Example

Find the value of the side indicated in the figure.

The two sides of interest are:

$$ADJ = 2"; \quad HYP = y$$

The ratio of ADJ/HYP is the cosine of the angle:

$$\cos(30°) = \frac{ADJ}{HYP} = \frac{2}{y}$$

From the plot (or using a scientific calculator), we see that when $x = 30°$, $\cos(x) = 0.87$.

$$\frac{2}{y} = 0.87 \rightarrow y = \frac{2}{0.87} = 2.3".$$

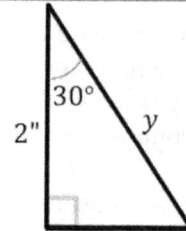

Right triangle for example.

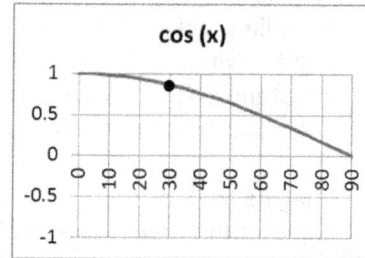

cos (x)

With these plots, you can also work backward to figure out the value of the angle from its trigonometric value. In other words, you can determine an unknown angle of a right triangle simply by knowing the measure of any two sides. This reverse operation is called taking the **inverse sine, cosine,** or **tangent.** It is also called taking the **arcsine, arccosine,** or **arctangent.**

Note on Using Scientific Calculators:

Calculators are great for giving accurate trigonometric values, but be sure you know if your calculator is set to degrees or radians for the angle, or you may get the wrong answer.

Term	Notation	Equation
Arcsine	$\text{asin}(x)$ or $\sin^{-1}(x)$	$x = \sin^{-1}\left[\frac{OPP}{HYP}\right]$
Arccosine	$\text{acos}(x)$ or $\cos^{-1}(x)$	$x = \cos^{-1}\left[\frac{ADJ}{HYP}\right]$
Arctangent	$\text{atan}(x)$ or $\tan^{-1}(x)$	$x = \tan^{-1}\left[\frac{OPP}{ADJ}\right]$

To do this in practice, simply take the ratio of two sides, locate that value on the y-axis of the correct plot, then find the corresponding x value.

Example

Find the value of the angle indicated in the figure.

We know two sides: $OPP = 2"; \ HYP = 3"$.

The ratio of OPP/HYP is the sine of the angle:

$$\sin(x) = \frac{OPP}{HYP} = \frac{2}{3} = 0.67$$

Taking the inverse, we say:

$$x = \sin^{-1}\left(\frac{2}{3}\right)$$

From the plot, when $\sin(x) = 0.67, x = 42°$.

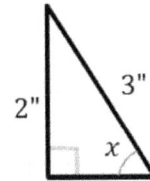

Right triangle for example.

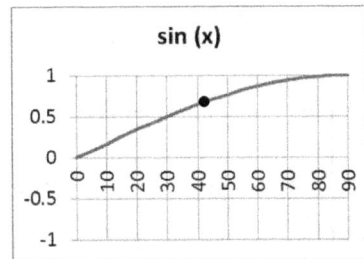

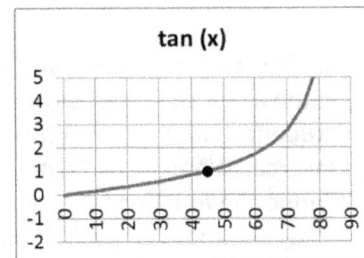

sin (x)

Example

Find the value of the angle indicated in the figure.

We know two sides: $ADJ = 2"; \ OPP = 2"$.

The ratio of OPP/ADJ is the tangent of the angle:

$$\tan(x) = \frac{OPP}{ADJ} = \frac{2}{2} = 1$$

Taking the inverse, we say:

$$x = \tan^{-1}(1)$$

Looking at the plot, when $\tan(x) = 1, x = 45°$.

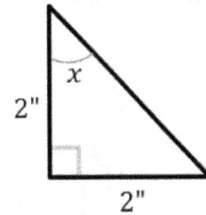

Right triangle for example.

tan (x)

There are other special ratios for a right triangle which can be used in a same manner, though they're not as popular. Also, once you know sine and cosine, you can derive the rest:

Term	Notation	Equation
Cosecant	$csc(x)$	$csc(x) = \dfrac{1}{sin(x)} = \dfrac{HYP}{OPP}$
Secant	$sec(x)$	$sec(x) = \dfrac{1}{cos(x)} = \dfrac{HYP}{ADJ}$
Cotangent	$cot(x)$	$cot(x) = \dfrac{1}{tan(x)}$ $= \dfrac{cos(x)}{sin(x)} = \dfrac{ADJ}{OPP}$

The plots for these ratios are given below.

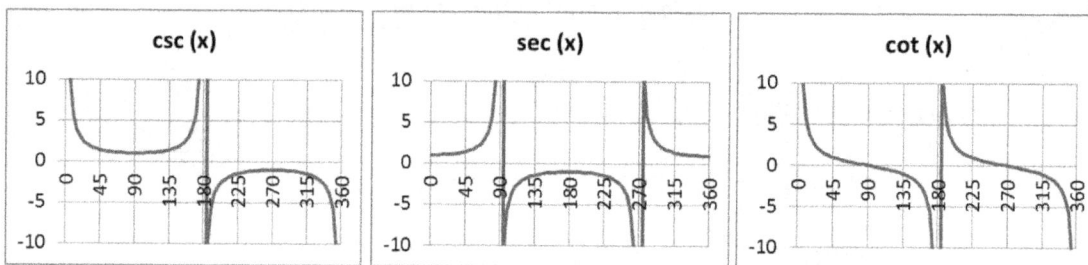

csc (x)

sec (x)

cot (x)

Taking inverses of cosecant, secant, and cotangent is the same as for the other trigonometric functions. Some students initially get confused by the "-1" notation for inverse. When placed over a function (like sine), it means "undo the function." When placed over an expression that gives a numerical result, it means "divide into 1."

$$sin^{-1}(x) \neq \frac{1}{sin(x)}$$

$$sin^{-1}(x) = asin(x) \qquad \text{gives an angle}$$

$$[sin(x)]^{-1} = \frac{1}{sin(x)} = csc(x) \qquad \text{gives a ratio}$$

Index

B-C

Other Books In This Series

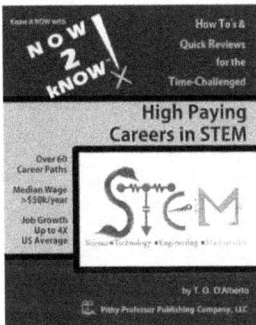

NOW 2 kNOW™ High Paying Careers in STEM

Plan for a career where your education dollars and hard work really pay off. STEM careers (science, technology, engineering, and math) are high paying, and companies are scrambling to find qualified candidates in the U.S. Better yet, the work is rewarding with opportunities to really impact the world!

NOW 2 kNOW™ Algebra 2 & Trigonometry

Expand on the concepts of Algebra I and Geometry with this two course text! Thorough and concise instruction coupled with over 200 problems and worked out solutions will have you on your way to Calculus in no time!

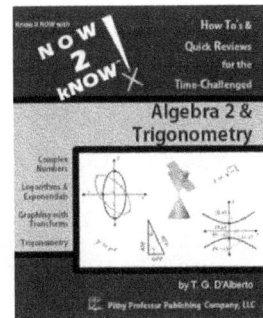

NOW 2 kNOW™ Calculus I

Calculus is the gateway to many financially stable and successful careers. You might think this would make it hard and inaccessible, but that simply isn't true. Calculus is actually very easy! Once you see the concepts outlined succinctly, you'll see how little it takes to become a master.

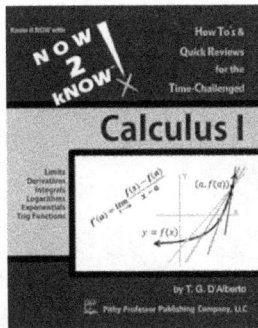

NOW 2 kNOW™ Calculus 2

Calculus 2 builds on the concepts of Calculus 1 with multi-variable functions and adds new concepts with infinite sequences and series. With thorough yet concise explanations and over 200 problems and worked out solutions, the NOW 2 kNOW™ Calculus 2 text makes learning math much easier!

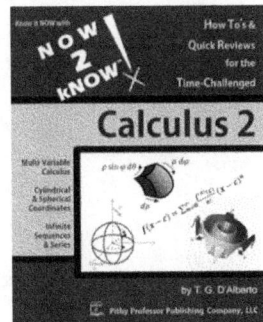

And for Review, see NOW 2 kNOW™ Algebra I!

Go to Amazon.com and Search "NOW 2 kNOW"
or visit www.NOW2kNOW.com for updates.

www.ingramcontent.com/pod-product-compliance
Lightning Source LLC
Chambersburg PA
CBHW05121520326
41519CB00025B/7127